Synthesis Lectures on Ocean Systems Engineering

Series Editor

Nikolas Xiros, University of New Orleans, New Orleans, LA, USA

The series publishes short books on state-of-the-art research and applications in related and interdependent areas of design, construction, maintenance and operation of marine vessels and structures as well as ocean and oceanic engineering.

C. Reid Nichols · Lynn Donelson Wright ·
Gary Zarillo

Integrated Coastal Resilience

 Springer

C. Reid Nichols
Marine Information Resources Corporation
Queenstown, MD, USA

Gary Zarillo
Ocean Engineering and Marine Sciences
Florida Institute of Technology
Melbourne, FL, USA

Lynn Donelson Wright
College of William and Mary
Virginia Institute of Marine Science
Gloucester Point, VA, USA

ISSN 2692-4420   ISSN 2692-4471 (electronic)
Synthesis Lectures on Ocean Systems Engineering
ISBN 978-3-031-68152-3   ISBN 978-3-031-68153-0 (eBook)
https://doi.org/10.1007/978-3-031-68153-0

This Springer imprint is published by the registered company Springer Nature Switzerland AG
The registered company address is: Gewerbestrasse 11, 6330 Cham, Switzerland

If disposing of this product, please recycle the paper.

Foreword

Integrated Coastal Resilience is an introductory geoscience book focusing on natural hazards, their impacts, and strategies for effective adaptations. Science issues addressed include relative sea level rise, storm damage, flooding, shoreline change, and ecological degradation. Compound effects such as flooding by combined storm surge, torrential rains, and river floods, as well as critical "tipping points" beyond which recovery is difficult, are also discussed. Not addressed in this book are specific emergency response measures or mitigation efforts that can be applied by decision-makers. However, this general reference provides insights into the data that can be accessed and evidence that can be applied to create the enabling capacity undergirding effective decision-making.

Coastal resilience is approached as a positive outcome achieved through collaboration by teams of diverse professionals (e.g., engineers, environmental scientists, health scientists, and social scientists). Natural hazards and data used to characterize extreme events are described. This book highlights information derived from federal observing networks and uses the Gulf of Mexico Coastal Ocean Observing System (GCOOS) as an example. The importance of citizen scientists who describe how recurring floods damage local structures is emphasized in this short textbook as well as how this information might influence policy makers. The way rising sea levels contribute to coastal change is discussed with numerous current references. For this reason, this general reference is particularly useful in helping engineering students to understand connections between the geosciences and environmental impacts on engineering projects.

This book provides information on data resources such as historical archives, in situ data, remotely sensed imagery, and numerical model output, which can be fused together as dashboards to benefit people living in coastal areas. Recent literature cited provides insights on environmental impacts ranging from recurring floods to shoreline protection. Data resources provided in this reference book and examples are useful in helping teachers to develop case studies that can help students translate basic knowledge into practice through real-world scenarios. The reader is apprised of the importance of big data, which includes examples from GCOOS and federal agencies including NOAA, NASA, and the US Army Corps of Engineers.

Integrated Coastal Resilience emphasizes the importance of collaboration and data fusion across temporal and spatial scales associated with varying environmental phenomena. This book puts engineering, geoscience, and social science concepts into a business context to improve preparation, resistance, adaptation, and recovery from natural hazards. State-of-the-art platforms and important monitoring networks that are used to collect meteorological and oceanographic data are introduced, along with the need to adopt data standards that provide actionable information. The complementary glossary that describes physical and social science terms related to coastal resilience is especially useful for all levels of environmental scientists, engineers, social scientists, and emergency managers.

Leonard J. Pietrafesa, Ph.D.
Professor Emeritus, North Carolina State University
Burroughs & Chapin Scholar
Coastal Carolina University
Ocean Isle Beach
North Carolina

Acknowledgments

Writing a general reference book on the topic of integrated coastal resilience would not have been possible without support from scientists and practicing engineers with experiences in multidisciplinary and integrated research projects. We especially appreciate insights from Lucie Cocquempot, Oceanographic Observation Coordinator at the French Research Institute, for Exploitation of the Sea (IFREMER) who ensured that this reference book touched on research that is improving coastal resilience in Europe. A detailed review by French Lt. Commander Amandine Kubié was valued, especially given her experience with marine engineering and sustainability. Dr. Olanrewaju S. Oladokun, a researcher at the Alfred Wegener Institute with experiences in marine technology, provided constructive feedback which was greatly appreciated. Edits and comments from Dr. Chris D'Elia, Professor Emeritus from Louisiana State University College of the Coast & Environment, improved the final product.

Contents

Introduction

The concepts of effective anticipation (prediction), resilience, and adaptation to the impacts of climate change on coastal communities are important to governments, industry, academia, communities, emergency managers and individuals. People tend to live near the coast for a variety of reasons, including the economic benefits derived from maritime commerce, marine fisheries, tourism, and recreation. Wright and Nichols (2019) described underlying coastal processes and coastal changes that are occurring worldwide. A compact literature update on the impacts of climate change on multiple coastal morphodynamic phenomena is offered by Wright and Thom (2023). The importance of resilience is evidenced by recent extreme weather events described by numerous types of organizations from insurance companies to national media (Frank, 2020; Lobo, 2021; Singha, 2019). Goodell (2017) described the threat of sea level rise to infrastructure, businesses, homes, and even cities. To mitigate the impact, government agencies have started to focus their attention on coastal resilience. Selected definitions by government agencies are listed in Table 1.1. Masselink and Lazarus (2019) concluded, "Coastal resilience is the capacity of the socioeconomic and natural systems in the coastal environment to cope with disturbances, induced by factors such as sea level rise, extreme events and human impacts, by adapting whilst maintaining their essential functions." The National Oceanic and Atmospheric Administration (NOAA) described coastal resilience as the ability of a community to "bounce back" after hazardous events such as hurricanes, coastal storms, and flooding (NOAA, 2022). The urgency has prompted organizations such as the U.S. Defense Advanced Research Projects Agency (DARPA) to develop the *Reefense* program, where collaborative researchers are developing hybrid biological and engineered reef-mimicking structures to help mitigate coastal flooding, erosion, and storm damage.

C. Reid Nichols et al., *Integrated Coastal Resilience*, Synthesis Lectures on Ocean Systems Engineering, https://doi.org/10.1007/978-3-031-68153-0_1

Table 1.1 Definitions of coastal resilience

Government/agency/study	Definition
European Commission (2020)	The ability not only to withstand and cope with challenges but also to undergo transitions, in a sustainable, fair, and democratic manner
Kazi et al. (2022)	The capacity of the socioeconomic and natural systems in the coastal environment to cope with disturbances, induced by both human and environmental factors, while adapting the essential functions of the systems to an improved state
Timmerman (1981)	Community-resilience measures, including both prevention and preparation, can reduce a disaster's impact. Resilience can be defined as the extent to which a community can withstand external shocks, including through emergency response and strategies to mitigate future harm
U.S. White House (Exec. Order No. 13653, 2013)	The ability to anticipate, prepare for and adapt to changing conditions and withstand and recover from disruptions

*Note*Based on these types of definitions the USACE considered prepare, absorb, recover, and adapt as four key principles for application (Rosati et al., 2015)

Because of climate change, sea level rise, altered river discharge, changes in the intensity and size of storms and changing demographics, coastal systems and communities of the future are seriously imperiled. Today, the threats to coasts presented by global climate change are orders of magnitude more severe than at any time in our history. Increases in the threats will require new engineering and management strategies to ensure future resilience. Meeting these emerging challenges will require major investments by nations, regions and communities along with unprecedented levels of transdisciplinary collaboration. A viable starting place involves effectively engaging stakeholders in resilience by fostering collaboration across disciplines, creating a shared understanding of the problems, promoting political acknowledgement of climate change and commitment to addressing the most pressing threats with funding and public policy, and implementing collaborative approaches to finding solutions. Transdisciplinary and transcultural collaborations are keys to creating and maintaining resilient coastal environments.

The sources of information on climate change are significant. Scientific reports on climate change have undergone tens of thousands of rigorous scientific peer reviews. The University Corporation for Atmospheric Research, the National Center for Atmospheric Research, The United Nations Environment Programme (UNEP) and the World Meteorological Organization (WMO) along with the NOAA Climate Prediction Center are among the numerous sources of data and model results on which global change projections are based. The Intergovernmental Panel on Climate Change (IPCC) is a foundational source of current scientific consensus on the causes and consequences of global change. Their sixth report, published in 2022, presents exhaustive and conclusive evidence that air temperatures at the earth's surface and sea temperatures are rising, leading to many other climate and ocean changes. Part 1 of the IPCC report is authored by 234 scientists from 66 nations and is underpinned by more than 14,000 peer reviewed scientific publications. It details the physical causes of climate change based on reliable data and information.

Part 2 of the Sixth IPCC report focuses on the anticipated impacts of predicted climate changes including draughts, famines, sea level rises, increased storminess, human health impacts, ecosystem degradation, and economic impacts. The latest draft report by the US Global Change Research Program concludes that the severity and risks of coastal hazards are increasing rapidly.

Here is a summary of what the expected future global changes include:

- *Atmospheric warming*–Global temperatures have risen 1 °C as greenhouse gases have trapped ever more heat from solar radiation. By the end of the century, the globally averaged temperature increase is expected to be somewhere between 1.8° and 4.0° Celsius (3.2° and 7.2° F). Most of the warming will be concentrated in the higher latitude and polar regions rather than in the tropics. The Arctic is warming more than twice as fast as the rest of the world.
- *Ocean warming*–Of the global warming that has taken place over the past two decades or so, 93% of the heat is currently estimated to be stored within the upper kilometer or so of the ocean. Ocean temperatures are expected to continue to rise significantly.
- *Storm intensity*–Tropical storm intensity is increasing, largely because of increased sea surface temperatures. The rapid intensification of many recent storms and the record number of hurricanes in 2020 and Hurricane Ian in 2022 have been attributed to unusually warm sea surface temperatures.
- *Evaporation and precipitation*–Rising temperatures are increasing the rates of evaporation, which are countered by increases in precipitation. As a result, draughts and floods are increasing.
- *Ice melting*–Melting of glacial ice in Alaska, Greenland, and Antarctica, along with the melting of permafrost is already taking place and expected to accelerate. The Arctic is warming nearly 3 times faster than elsewhere on earth.
- *Global Sea level*–is rising and will continue to do so throughout the twenty-first century. Sea level is expected to be significantly higher than today by the end of the century. The latest study by the US Global Change Research Program shows that sea levels by 2050 could be 1.5–2.0 ft higher than today.
- *Altered ocean circulation*–Major, large scale ocean current systems, such as the Gulf Stream, are *slowing down*. Reduction of Gulf Stream transport causes coastal ocean sea level to rise along the US east coast at rates faster than the global trend

Knowledge of climate change factors, especially in the dynamic coastal zone, is important for the design and build of resilient structures and facilities. Modern day coastlines are associated with increases in population density and economic activity (Neuman et al., 2015). Engineers must consider how climate change will impact the effectiveness of structures from docks and coastal roads to offshore buoys and offshore platforms, which are influenced, directly or indirectly, by waves, tides, and shallow water processes. Climatic factors include changing weather patterns and the resultant environmental loads from snow, ice, wind, waves, and currents (DNV, 2019; Nichols & Raghukumar, 2020). Organizations such as the European Union, IPCC, NOAA, U.S. Army Corps of Engineers (USACE), and WMO help to identify climate change trends, which supports the design of resilient structures and communities.

Improving coastal resilience is an ongoing imperative to cope with the consequences of extreme weather events (e.g., flooding, heat waves, heavy downpours, hurricanes, and wildfires). More frequent extremes are driven by a warming atmosphere. Scientists working on the IPCC pointed out that the intensity and frequency of extreme weather and climate events resulting from climate change will greatly impact human, social, and economic systems (Parry et al., 2007). The latest report by the IPCC focused on the anticipated impacts of predicted climate changes and the increasing severity and risks of coastal hazards (IPCC, 2022). Researchers and journalists from around the globe (e.g., Francis, 2019; Ge et al., 2022; Golub et al., 2022; Vaughan, 2020) have provided evidence that extreme weather events are becoming more frequent or intense owing to global warming. Climate change has also been attributed to the slowing of the Gulf Stream, which contributes to rising sea levels along the east coast of the United States (Ezer & Atkinson, 2017). Resilience planning by governments and industries needs to account for chronic events like rising sea levels, worsening air quality, and population migrations.

Integrated coastal resilience implies more than just designing structures to absorb or avoid damage without suffering complete failure. Resilience also entails taking the necessary actions to ensure food, energy, and water security (Vorosmarty et al., 2023). Linkov et al. (2014) described a systematic approach to resilience while highlighting the value of stakeholder management to ensure the timely and broad acceptance of resilience concepts. Implementing capabilities that improve resilience involves identifying, prioritizing, and engaging key people and organizations (Allen et al., 2021). Intermediary or broker organizations could be used to support collaboration among stakeholders from different disciplines, organizations, and places (Stadtler & Karakulak, 2020; Stadtler & Probst, 2012). Owing to the complexity of making a community more resilient, a neutral third party can facilitate the sharing of resources, especially new technologies and communication tools. Nichols and Wright (2020) provided some history on collaborative research managed by the Southeastern Universities Research Association, an intermediary organization, that resulted in a series of cross-disciplinary coastal resilience workshops and development of the Coastal and Ocean Modeling Testbed for NOAA. Example roles for an intermediary involve engaging stakeholders, facilitating understanding of

Table 1.2 Selected example environmental, engineering, and community factors that impact coastal resilience

Environmental	Engineering	Community
Climate (norms, means, & extremes)	Drainage	Affordable housing
Coast type (barrier island, delta, estuary, mangrove, etc.)	Environmental loads	Age
	Flood plain planning	Education
Geomorphology	Maintenance	Ideology and perspective
Storm surge	Off-grid energy	Inundation
Water level fluctuations	Real-property assets	Pollution
Wave climate	Risk	Population density
Wave slamming	Sewage backflow prevention	Water quality
Wind speed and direction		Wealth

requirements, objectively testing new modelling and simulation techniques, demonstrating innovations for sponsors, technology transfer, and effecting communications through stakeholder workshops. Achieving integrated coastal resilience requires well-planned collaboration among government, university, industry, and community partners. In extreme cases, collaboration among government, university, and community stakeholders might lead to data-driven decisions that involve the relocation or abandonment of coastal communities (Fiorentino et al., 2023; Pinter, 2021; Taarup-Esbensen, 2022). Integrated coastal resilience brings together multidisciplinary professionals and the community to protect property and save lives.

An integrative, holistic approach that includes multiple disciplines to address the environmental, engineering, and community factors is fundamental to success. Kim et al. (2022) indicated the relevance and challenges behind managing these transdisciplinary research projects. Example factors that are essential to consider together are listed in Table 1.2. Biologists have established resilience concepts that describe how complex biological systems adapt and recover from disturbances such as fires, floods, and even oil spills (Gunderson et al., 2009; Holling, 1973). Numerous physical scientists are meeting the challenge of recurrent flooding through advances such as street-level modeling (Blumberg et al., 2015; Liu et al., 2020; Wang et al., 2014). Scientists and engineers have studied storm damages to assess the way natural infrastructure, built infrastructure, and combinations of natural and built infrastructure enhance coastal resilience by providing important storm and coastal flooding protection (Sutton-Grier et al., 2015). From the community perspective, Buchanan et al. (2020) discussed the impact of sea level rise on affordable housing. An integrated approach puts these research examples together to address local issues, which is key since factors such as sea level rise are not uniform across the globe. Successful projects include people and organizations who think systemically.

Episodic events such as earthquakes, hurricanes, and floods often spur research thrusts aimed at developing innovative engineering techniques, more effective emergency response procedures, and new laws. Following Hurricanes Katrina in 2005 and Sandy in 2012, organizations such as the Environmental Protection Agency (EPA), NOAA, and the USACE provided many valuable resources to think resiliently (Touzinsky et al., 2016). The general approach to resiliency has involved knowing what hazards are possible in an area of interest, assessing vulnerability and risks, investigating options, prioritizing and planning, and acting. The conceptual model that the USACE and others have applied is illustrated in Fig. 1.1. Table 1.3 lists some organizations and programs that are trying to improve coastal resilience. Non-Governmental Organizations (NGOs) such as The Nature Conservancy (TNC) and Climate Central have provided analytical tools that allow citizen scientists to leverage big data within their local communities through capabilities such as the Resilient Land Mapping Tool (see https://maps.tnc.org/resilientland/) and Surging Seas Risk Finder and Map (see https://sealevel.climatecentral.org/).

Geoscientists, engineers, and social scientists consider resilience as the ability to absorb or avoid damage without suffering complete failure for ecosystems, structures, industries, and communities. Whereas there are numerous theories related to resilient engineering (Righi et al., 2015), this book looks at approaches that consider environmental, engineering, and community factors together. The integration of resilience concepts provides a larger combination of knowledge, skills, tools and resources for people to continually assess and adjust to the dynamic coastal environment. The collective of people interested in improving resilience can access knowledge, improve problem-solving, and speed up product development through effective collaboration.

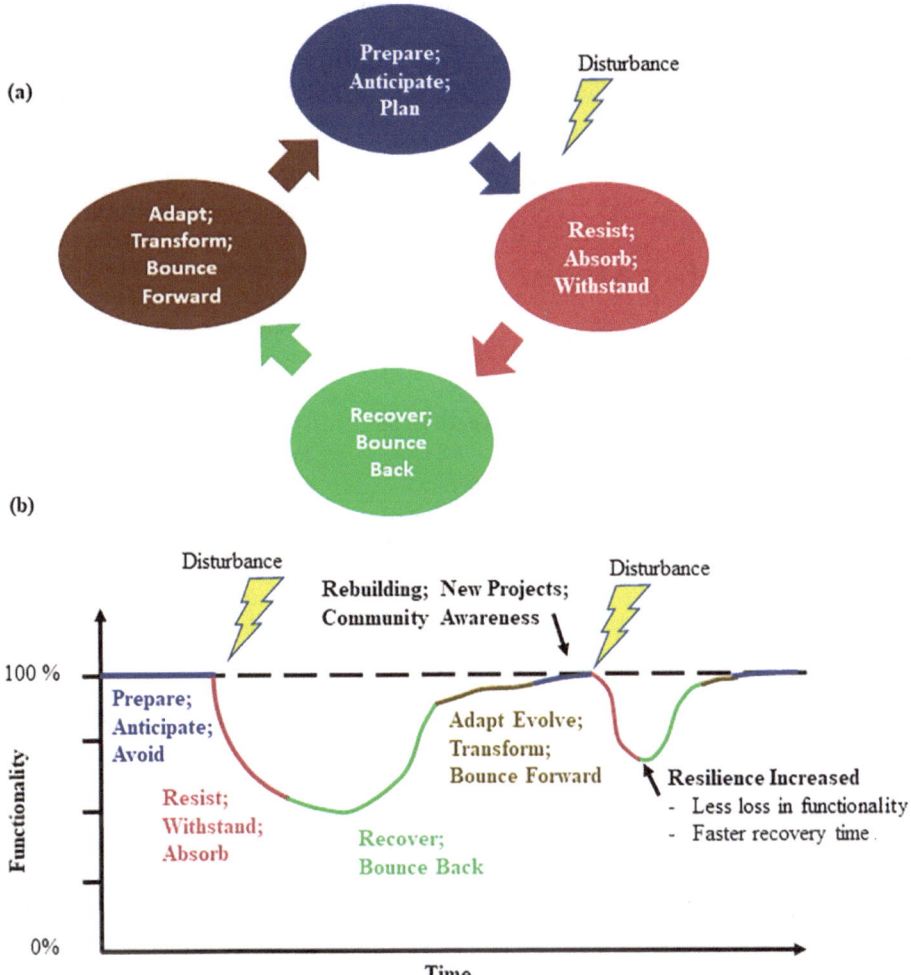

Fig. 1.1 Coastal resilience concepts. **a** Resilience involves an ongoing cycle that is activated by a disturbance. **b** The four principles of resilience are put into context before and after a series of disturbances. Owing to resilience, the impact from a disturbance is reduced and recovery is faster following subsequent disturbances. This figure was copied from Rosati et al. (2015). Elements of the figure have been described by Linkov et al. (2014); Schultz et al. (2012) and Touzinsky et al. (2016)

Table 1.3 Example coastal resilience programs and advances

Sponsor	Selected programs	Example advances or evidence
DARPA (https://www.darpa.mil/)	Reefense	DARPA has selected investigators from Rutgers University, University of Hawaii, and University of Miami to lead efforts to mitigate damages due to storm surge and flooding through the development and demonstration of self-healing, hybrid biological, and engineered reef-mimicking structures (e.g., Saal, 2023)
NOAA (https://www.noaa.gov/)	Coastal Resilience Grant Program; National Coastal Resilience Fund; Transformational Habitat Restoration and Coastal Resilience Grants	TNC received funding to restore coastal habitat in the He'eia watershed and Kāne'ohe Bay, Oahu, Hawaii. The project involved invasive species removal, native species replanting, and community education (Campbell & Campbell, 2017). The effort was intended to strengthen ecological and community resilience to extreme weather events
National Science Foundation (NSF, https://www.nsf.gov/)	Convergence accelerator; Resilient & Intelligent NextG systems; Risk & Resilience	Northwestern University along with its academic and government partners have been funded to complete a project called "Systems Approaches for Vulnerable Evaluation and Urban Resilience." The project merges natural science, social science, data science and engineering to predict extreme events such as heat waves, air quality and flooding, as well as assess vulnerabilities within the Chicago metropolitan area (Crisologo et al., 2019)

(continued)

Table 1.3 (continued)

Sponsor	Selected programs	Example advances or evidence
Resilient Asia program (https://www.worldbank.org/en/region/sar/brief/resilient-asia-program)	Strengthen early warning systems and climate resilient investments in shoreline protection for atoll states	Nakayama et al. (2022) investigated adaptations to climate change ranging from migration to land reclamation and raising to the development of floating platforms for atoll states such as the Republic of the Marshall Islands. Crameri and Ellison (2024) highlighted the importance of mangroves to mitigate sea level rise vulnerability on Marshall Islands atolls
Strategic Environmental Research and Development Program (SERDP; https://www.serdp-estcp.org/)	Regional sea Level scenarios for coastal risk management: Managing the uncertainty of future sea level Change and extreme water levels for Department of defense coastal sites Worldwide	SERDP has sponsored several studies to assess the vulnerability and potential impacts from climate change at certain US military bases and stations (Hall et al., 2016; Parris et al., 2012). The development of sea level and extreme water level scenarios and an accompanying database is available online at https://drsl.serdp-estcp.org/
United Nations, Global Ocean Observing System (GOOS; https://goosocean.org/)	The storm surge exemplar project	The ocean observing co-design program is evolving an integrated ocean observing system to cope with issues such as storm surge. The Euro-Mediterranean Center on Climate Change (CMCC) is leading the Storm Surge Exemplar Project for GOOS to utilize observations and models to understand the behavior of the ocean and to predict changes in precipitation and storm surge owing to climatic changes (Accarino et al., 2021; Biggins, 2004)

References

Allen, T., Behr, J., Bukvic, A., Calder, R. S. D., Caruson, K., Connor, C., D'Elia, C., Dismukes, D., Ersing, R., Franklin, R., Goldstein, J., Goodall, J., Hemmerling, S., Irish, J., Lazarus, S., Loftis, D., Luther, M., McCallister, L., McGlathery, K., … Zinnert, J. C. (2021). Anticipating and adapting to the future Impacts of climate change on the health, security and welfare of Low Elevation Coastal Zone (LECZ) communities in Southeastern USA. *Journal of Marine Science and Engineering, 9*, 1196. https://doi.org/10.3390/jmse9111196

Accarino, G., Chiarelli, M., Fiore, S., Federico, I., Causio, S., Coppini, G., & Aloisio, G. (2021). A multi-model architecture based on long short-term memory neural networks for multi-step sea level forecasting. *Future Generation Computer Systems, 124*, 1–9. https://doi.org/10.1016/j.fut ure.2021.05.008

Blumberg, A. F., Georgas, N., Yin, L., Herrington, T. O., & Orton, P. M. (2015). Street-scale modeling of storm surge inundation along the New Jersey Hudson River waterfront. *Journal of Atmospheric Technology, 32*(8), 1486–1497. https://doi.org/10.1175/JTECH-D-14-00213.1

Biggin, S. (2004). Italy hosts a climate research center. *Science, 306*(5705), 2171. https://www.sci ence.org/doi/10.1126/science.306.5705.2171c

Buchanan, M. K., Kulp, S., Cushing, L., Morello-Frosch, R., Nedwick, T., & Strauss, B. (2020). Sea level rise and coastal flooding threaten affordable housing. *Environmental Research Letters, 15*(12), Article 124020. https://doi.org/10.1088/1748-9326/abb266

Campbell, H. V., & Campbell, A. M. (2017). Community-based watershed restoration in Heʻeia (Heʻeia ahupuaʻa), Oʻahu, Hawaiian Islands. *Case Studies in the Environment, 1*(1), 1–8. https://doi.org/10.1525/cse.2017.sc.450585

Crameri, N. J., & Ellison, J. C. (2024). Atoll mangrove progradation patterns: Analysis from Jaluit in the Marshall Islands. *Estuaries and Coasts, 47*, 935–948. https://doi.org/10.1007/s12237-024-01331-0

Crisologo, I., Luo, H., Medendorp, A., Garcia, M. H., Collis, S. M., & Horton, D. E. (2019). *Using high-resolution radar rainfall products to improve city-scale flood models for urban resilience [Poster presentation].* American Geophysical Union 2019 Fall Meeting. San Francisco, CA. https://doi.org/10.1002/essoar.10501802.1

DNV. (2019). *Environmental conditions and environmental loads*, DNV-RP-C205. Det Norske Veritas

European Commission. (2020). *Strategic foresight: Charting the course towards a more resilient Europe.* 2020 Strategic foresight report. https://eur-lex.europa.eu/legal-content/EN/TXT/PDF/?uri=CELEX:52020DC0493&from=EN

Exec. Order No. 13653, 78 FR 66817, pp. 66817–66824. (2013). https://www.govinfo.gov/content/pkg/FR-2013-11-06/pdf/2013-26785.pdf

Ezer, T., & Atkinson, L. P. (2017). On the predictability of high water level along the U.S. East Coast: Can the Florida Current measurement be an indicator for flooding caused by remote forcing? *Ocean Dynamics.* https://doi.org/10.1007/s10236-017-1057-0

Fiorentino, S., Sielker, F. & Tomaney, J. (2023). Coastal towns as 'left-behind places': Economy, environment and planning. *Cambridge Journal of Regions, Economy and Society, 17*(1), 103–116. https://doi.org/10.1093/cjres/rsad045

Francis, J. (2019). Rough weather ahead. *Scientific American., 320*(6), 46–53.

Frank, T. (2020). With sea level rise, high-tide flooding spikes along U.S. coasts. *Scientific American.* https://www.scientificamerican.com/article/with-sea-level-rise-high-tide-flooding-spi kes-along-u-s-coasts/

Ge, Z.-A., Chen, L., Li, T., & Wang, L. (2022). How frequently will the persistent heavy rainfall over the middle and lower Yangtze River basin in summer 2020 happen under global warming? *Advances in Atmospheric Sciences, 39*(10), 1673–1692. https://doi.org/10.1007/s00376-022-1351-8

Golub, A., Govorukha, K., Mayer, P., & Rübbelke, D. (2022). Climate change and the vulnerability of Germany's power sector to heat and drought. *Energy Journal., 43*(3), 157–183. https://doi.org/10.5547/01956574.43.3.agol

Goodell, J. (2017). *The water will come: Rising seas, sinking cities, and the remaking of the civilized world*. Little, Brown and Company.

Gunderson, L. H., Allen, C. R., & Holling, C. S. (Eds.) (2009). *Foundations of Ecological Resilience*. Island Press

Hall, J. A., Gill, S., Obeysekera, J., Sweet, W., Knuuti, K., & J. Marburger, J. (2016). *Regional sea level scenarios for coastal risk management: Managing the uncertainty of future sea level change and extreme water levels for department of defense coastal sites worldwide*. U.S. Department of Defense, Strategic Environmental Research and Development Program. https://drsl.serdp-estcp.org/Docs/CARSWG_SLR.pdf

Holling, C. S. (1973). Resilience and stability of ecological systems. *Annual Review of Ecology and Systematics., 4*, 1–23. https://doi.org/10.1146/annurev.es.04.110173.000245

IPCC. (2022). *Climate Change 2022: Impacts, adaptation and vulnerability. Contribution of working group II to the sixth assessment report of the intergovernmental panel on climate change.* Cambridge University Press. https://www.ipcc.ch/report/ar6/wg2/downloads/report/IPCC_AR6_WGII_FrontMatter.pdf

Kazi, S., Urrutia, I., van Ledden, M., Laboyrie, J. H., Verschuur, J., Khan, Z-u, H., Jongejan, R., Lendering, K., & Mancheño, A., J. (2022). *Bangladesh: Enhancing coastal resilience in a changing climate*. The World Bank. https://documents1.worldbank.org/curated/en/099552209012279085/pdf/IDU0c262e78c0311a04ee80be0a03c2ce4024c98.pdf

Kim, K. M., Douglas, M. M., Pannell, D., Setterfield, S. A., Hill, R., Laborde, S., Perrott, L., Álvarez-Romero, J. G., Beesley, L., Canham, C., & Brecknell, A. (2022). When to use transdisciplinary approaches for environmental research. *Frontiers in Environmental Science, 10*, 840569. https://doi.org/10.3389/fenvs.2022.840569

Linkov, I., Bridges, T., Creutzig, F., Decker, J., Fox-Lent, C., Kröger, W., Lambert, J. H., Levermann, A., Montreuil, B., Nathwani, J., Nyer, R., Renn, O., Scharte, B., Scheffler, A., Schreurs, M., & Thiel-Clemen, T. (2014). Changing the resilience paradigm. *Nature Climate Change, 4*, 407–409. https://doi.org/10.1038/nclimate2227

Liu, Z., Wang, H., Zhang, Y. J., Magnusson, L., Loftis, J. D., & Forrest, D. (2020). Cross-scale modeling of storm surge, tide, and inundation in Mid-Atlantic Bight and New York City during Hurricane Sandy, 2012. *Estuarine, Coastal and Shelf Science, 233*, Article 106544. https://doi.org/10.1016/j.ecss.2019.106544

Lobo, N. V. (2021). *Extreme heat in Western and Southern Europe: A real and increasingly regular peril of climate change*. Swiss Re Group. https://www.swissre.com/risk-knowledge/mitigating-climate-risk/extreme-heat-in-western-southern-europe.html

Masselink, G., & Lazarus, E. D. (2019). Defining coastal resilience. *Water, 11*(12), 2587. https://doi.org/10.3390/w11122587

Nakayama, M., Fujikura, R., Okuda, R., Fujii, M., Takashima, R., Murakawa, T., Sakai, E., & Iwama, H. (2022). Alternatives for the Marshall Islands to cope with the anticipated sea level rise by climate change. *Journal of Disaster Research, 17*(3), 315–326. https://doi.org/10.20965/jdr.2022.p0315

Neumann, B., Vafeidis, A. T., Zimmermann, J., & Nicholls, R. J. (2015). Future coastal population growth and exposure to sea-level rise and coastal flooding-A global assessment. *PLoS ONE, 10*(3), e0118571. https://doi.org/10.1371/journal.pone.0118571

Nichols, C. R., & Wright, L. D. (2020). The evolution and outcomes of a collaborative testbed for predicting coastal threats. *Journal of Marine Science and Engineering, 8*(8), 612. https://doi.org/10.3390/jmse8080612

Nichols, C. R. & Raghukumar, K. (2020). *Marine environmental characterization.* Springer Cham. https://doi.org/10.1007/978-3-031-02490-0

NOAA. (2022). *What is resilience?* National Ocean Service website. https://oceanservice.noaa.gov/facts/resilience.html

Parris, A. S., Bromirski, P., Burkett, V., Canyan, D. R., Culvert, M. E., Hall, J., Horton, R. M., Knuuti, K., Moss, R. H., Obeysekera, J., Sallenger, A. H., & Weiss, J. (2012). *Global sea level rise scenarios for the US National Climate Assessment. NOAA Technical Memorandum OAR CPO-1.* Department of Commerce, Climate Program Office. https://repository.library.noaa.gov/view/noaa/11124

Parry, M., Cansiani, O., Palutikof, J., van der Linden, P., & Hanson, C. (Eds.) (2007). *Climate change 2007: Impacts, adaptation, and vulnerability.* Cambridge University Press. https://www.ipcc.ch/site/assets/uploads/2018/03/ar4_wg2_full_report.pdf

Pinter, N. (2021). The lost history of managed retreat and community relocation in the United States. *Elementa: Science of the Anthropocene, 9*(1). https://doi.org/10.1525/elementa.2021.00036

Righi, A. W., Saurin, T. A., & Wachs, P. (2015). A systematic literature review of resilience engineering: Research areas and a research agenda proposal. *Reliability Engineering & System Safety, 141*, 142–152. https://doi.org/10.1016/j.ress.2015.03.007

Rosati, J. D., Touzinsky, K. F., & Lillycrop, W. J. (2015). Quantifying coastal system resilience for the US Army Corps of Engineers. *Environment Systems and Decisions, 35*, 196–208. https://doi.org/10.1007/s10669-015-9548-3

Stadtler, L., & Karakulak, Ö. (2020). Broker organizations to facilitate cross-sector collaboration: At the crossroad of strengthening and weakening effects. *Public Administration Review, 80*(3), 360–380. https://doi.org/10.1111/puar.13174

Stadtler, L., & Probst, G. (2012). How broker organizations can facilitate public-private partnerships for development. *European Management Journal, 30*(1), 32–46. https://doi.org/10.1016/j.emj.2011.10.002

Singha, M. (2019). *Odisha government to tweak its climate action plan after alarming crisis.* The Times of India. https://timesofindia.indiatimes.com/city/bhubaneswar/odisha-government-to-tweak-its-climate-action-plan-after-alarming-crisis/articleshow/71842693.cms

Saal, K. (2023). 'Reefense' restoration project: Developing hybrid reefs to reduce wave energy and shield coastal communities: Oyster habitats could play an important role in protecting the U.S. Gulf Coast—and eventually other coastlines—from damaging waves and storm damage. *Stormwater, 24*(4), 28–30

Schultz, M. T., McKay S. K., & Hales L. Z. (2012). *The quantification and evolution of resilience in integrated coastal systems.* U.S. Army Corps of Engineers, ERDC TR-12-7. https://el.erdc.usace.army.mil/elpubs/pdf/tr12-7.pdf

Sutton-Grier, A. E., Wowk, K., & Bamford, H. (2015). Future of our coasts: The potential for natural and hybrid infrastructure to enhance the resilience of our coastal communities, economies and ecosystems. *Environmental Science & Policy, 51*, 137–148. https://doi.org/10.1016/j.envsci.2015.04.006

Taarup-Esbensen, J. (2022). Community resilience—systems and approaches in remote settlements. *Progress in Disaster Science, 16*, 100253. https://doi.org/10.1016/j.pdisas.2022.100253

Timmerman, P. (1981). *Vulnerability, resilience, and the collapse of society, Environmental Monograph No. 1, Institute of Environmental Studies.* University of Toronto

Touzinsky, K., Rosati, J., Fox-Lent, C., Becker, A., & Luscher, A. (2016). Advancing coastal systems resilience research: Improving quantification tools through community feedback. *Shore & Beach, 84*(4), 30–37.

Vaughan, A. (2020). A whirlwind of extreme weather. *New Scientist, 248*(3313/3314), 28–29.

Vörösmarty, C. J., Melillo, J. M., Wuebbles, D. J., Jain, A. K., Ando, A. W., Chen, M., Tuler, S., Smith, R., Kicklighter, D., Corsi, F., Fekete, B., Miara, A., Bokhari, H. H., Chang, J., Lin, T.-S., Maxfield, N., Sanyal, S., & Zhang, J. (2023). Applying the framework to study climate induced Extremes on food, energy, and water systems (C-FEWS): The role of engineered and natural infrastructures, technology, and environmental management in the United States Northeast and Midwest. *Frontiers in Environmental Science, 11*, 1070144. https://doi.org/10.3389/fenvs.2023.1070144

Wright, L. D., & Nichols, C. R. (Eds.) (2019). *Tomorrow's coasts: complex and impermanent, coastal research library*, 27, Springer.

Wright, L. D., & Thom, B. G. (2023). Coastal morphodynamics and climate change: a review of recent advances. *Journal of Marine Science and Engineering, 11*(10), 1997. https://doi.org/10.3390/jmse11101997

Wang, H., Loftis, J. D., Liu, Z., Forrest, D., & Zhang, J. (2014). Storm surge and sub-grid inundation modeling in New York city during hurricane sandy. *Journal of Marine Science and Engineering, 2*(1), 226–246. https://doi.org/10.3390/jmse2010226

Natural Hazards

Natural hazards include many geophysical and biological phenomenon that have negative effects on humans, animals, and ecosystems. These complex phenomena cannot be completely understood by the lone researcher. Collaborative research approaches benefit from the integration of knowledge, methods, and expertise from across science and engineering disciplines. For this reason, natural hazards are studied by multidisciplinary and integrated teams of scientists as illustrated in Table 2.1. These scientists and science journalists are increasingly on the front lines of reporting the causes and impacts of extreme events and the need for improved coastal resilience. Lund (2015) highlighted the value and progress made to improving the performance of water management systems by projects that considered fundamentals from both physical and social sciences. From a community perspective, in the face of coastal hazards, public administrators must manage the emergency while dealing with numerous social factors such as the ideologies and morays of the local population, available resources, electoral politics, the politics of blame, and the political rewards for preparedness and relief (Mulligan et al., 2019). Psychologists and social workers have important roles to perform such as psychosocial counseling and making relief services available (Bauwens & Naturale, 2017). Coastal resilience requires understanding and effective and timely response to natural hazards, which can only be realized through communication, coordination, cooperation, and collaboration across organizations and disciplines. The way these 4 Cs are applied to coastal resilience is detailed in Table 2.2.

15
C. Reid Nichols et al., *Integrated Coastal Resilience*, Synthesis Lectures on Ocean Systems Engineering, https://doi.org/10.1007/978-3-031-68153-0_2

Table 2.1 General roles and responsibilities of natural hazard investigators

Scientists & engineers	Example work areas
Biologists	Determining causes of harmful algal blooms (HABs), dense formations of cyanobacteria, and locations of dead zones
Civil engineers	Assessing risks, managing threats, and mitigating impacts of natural hazards on the built environment
Environmental lawyers	Researching and preparing legal cases, interviewing various people involved in the case as well as experts in the field, such as scientists and engineers, to help substantiate the case. Environmental laws include federal and state statutes that concern wildlife, endangered species, habitats, public lands, logging and forestry, natural resources, hazardous and toxic wastes, air and water pollution, and other environment-related matters
Environmental scientists	Protecting and preserving the environment by drawing on the natural sciences to examine human impacts on the environment
Geographers	Characterizing intensity and pattern of loss distribution from natural hazards based on geographical factors such as population density
Geologists	Studying the effects of earthquakes, subsidence, volcanic eruptions, and initiating events for tsunamis
Hydrologists	Understanding the driving forces for droughts and floods and related events such as deposition, landslides, and river scour
Mechanical, ocean, and structural engineers	Computing environmental loads, structural response to those loads, and measuring damage of structural systems, non-structural components, contents, and equipment following extreme events
Meteorologists	Addressing challenges posed by atmospheric hazards such as climate change, extreme weather, and volcanic plumes
Oceanographers	Measuring changing current patterns, ocean chemistry, water level fluctuations such as tides, waves, and initiating events for recurring floods, sea level rise, and storm surges. Tracking dangerous marine life such as harmful algal blooms

(continued)

Table 2.1 (continued)

Scientists & engineers	Example work areas
Psychologists/Social workers	Investigating and treating mental health consequences for victims of natural hazards such as Post Traumatic Stress Disorder
Public administrators	Managing the politics that shape the management of natural hazards, e.g., electoral politics, emergency management, and the political benefits associated with disaster relief spending
Public health scientists	Promoting health and disease prevention by examining medical data, performing tests on lab samples, and analyzing test results
Science reporters	Writing scientific articles related to natural hazards for general circulation magazines, science magazines geared to the general public, magazines for scientists and engineers, and newspapers

Note There are overlapping roles and responsibilities among the scientists and engineers who are involved in coping with natural hazards. Owing to the spatial and temporal variability of data, most scientists complete investigations by applying mapping and modeling techniques. Improvement needs to be made to effectively link the social and physical scientists. Breslauer and Breslauer (2023) discussed similarities in the nature of physical and social sciences from a system complexity perspective

2.1 Collaborative Processes and Decision Support Systems

Collaboration to improve coastal resilience is essential since evidence of climate change has been documented by many observing systems that comprise the GOOS. Liu et al. (2015) defined the installation and operation of these systems in relation to geopolitical and environmental considerations. Observations by satellite altimeters and measurements by water level gages have recorded a rise in sea level. Weather stations have documented droughts and heat waves. Networked stream gage stations that measure parameters such as water depths to a reference height and water velocity, have observed floods and water shortages. Emergency responders use these observations (data) in planning and during their operations. These data are also used to assess impacts, including effects on human life. Organizations such as NOAA ensure the availability of databases, imagery, time series, and forecasts so that decision tools can be developed that support coastal resilience, climate change adaptation, and inform stewardship and resource management. Collaboration among scientists, engineers, and citizens allows us to combine diverse perspectives, skills, and experiences.

Collaborative research and the development of capacity by organizations that work with the WMO is evidenced through development of the WMO Integrated Global Observing System (WIGOS). Such programs are focused on the collection of quality data and the

Table 2.2 The 4Cs to understand and respond to natural hazards

Partner interactions	Brief definition	Examples
Communication	Exchanging ideas and information	Sharing information so that all stakeholders can thrive during a disturbance. Effective communications include scenario-based stakeholder workshops, technology exchange meetings, and publications. Since communication methods are constantly evolving, it's imperative to leverage the latest technologies. Social and behavior change communication might apply communication approaches such as mass media, social media, digital communication, community-level activities, interpersonal communication and advocacy to influence social norms and behaviors (see https://www.centreforsbcc.org/what-is-sbcc/)
Coordination	Sharing vision and goals	Tracing requirements to user's needs and managing stakeholders to succeed in implementing policies that reduce human suffering while promoting resilience. Align and synchronize to create efficiency. Coordination allows stakeholders who are working on different parts of a project to know how their parts fit together. Lanier et al. (2018) identified strategies and tools that support team communication and coordination
Cooperation	Aligning goals and objects; co-design	Thinking systemically to properly share goals while trying to avoid unintended consequences. Implementing processes that facilitate execution and accountability. Working with end-users to ensure effective products that meet partner goals. Testing and evaluating how well resilience approaches work for stakeholders. Castañer and Oliveira (2020) defined cooperative behavior as actions undertaken by the partners to achieve the collectively envisioned goal
Collaboration	Co-creating with multiple stakeholders and organizations	Allowing all stakeholders to participate in resilience efforts, whether they are internal or external. Sharing resources, analysis, and support to process and overcome hardships. Including under-represented groups to build community and a resilient culture. Hückstädt (2023) reported that collaboration success is influenced by factors such as shared goals, commitment, cooperative climate, continuous coordination, and clear communications

improvement of extreme weather forecasts since many predicted impacts from natural hazards are inaccurate. For example, European countries were hit by basically unanticipated extreme floods during 2021 (Lehmkuhl et al., 2022) and drought conditions that rendered some rivers non-navigable during 2022 (European Commission, 2022). To better protect property and save lives, governments, universities, and industry are all involved in efforts to assess the risks of hazards and to develop plans that reduce vulnerability. Programs such as WIGOS are focused on capturing additional requirements on traditional weather fore-casting and warning services owing to the impacts of climate change that include sea level rise, increased frequency of various extreme weather and climate events, and geographic shifts in major agricultural growing zones (WMO, 2019). WIGOS meets these require-ments by contributing observations and tools to assist communities in the reduction of disaster risks and in adapting to climate changes. In accordance with Ocean Best Prac-tices (see https://www.oceanbestpractices.org/), WIGOS resources include a framework that includes public web-based analysis tools.

Climate change is disproportionately affecting developing countries, such as flooding in Papua New Guinea during monsoons and tsunamis (World Bank Group, 2021) and the destruction of low-income dwellings within all countries including communities such as Fort Myers[1] in Florida during Hurricane Ian (SECOORA, 2022). Developing countries may be at risk owing to poor infrastructure, limited resources, and unstable institutions. Morea and Samanta (2020) described the use of imagery and Geographic Information Systems (GIS) to assess flood risk zones in threatened areas in Papua New Guinea[2] and planning for both structural and non-structural solutions to recurring floods. Regional Associations such as SECOORA (Southeast Coastal Ocean Observing Regional Associa-tion) are part of the U.S. Integrated Ocean Observing System (IOOS) and provide users with various types of current and historic data. These data are essential for planning by personnel from coastal engineers to city managers. The combination of historical informa-tion and imagery, in situ data from sensors, and numerical model output can be integrated using GIS to develop decision-aid products such as forecast charts and wave atlases.

Collaboration among government, university, and industry partners improves the like-lihood of success with resilience improvements, especially given the lack of scientific certainty on the timing and spatial extent of many phenomena. Palmer and Stevens (2019) discussed uncertainty and opportunities to use supercomputers to reliably simu-late extremes. Adaptation involves developing and integrating risk reduction strategies, e.g., those developed by environmentalists, engineers, and communities. Toward this, in many coastal regions across the globe, scientists are developing climate risk assessments, public administrators are assessing priorities and planning follow-up to resist the threat of extremes, educators are raising awareness, governments are establishing partnerships that intermediary organizations often manage, and researchers are highlighting vulnera-bilities across disciplines. These activities are important since they benefit other affected coastal communities through the sharing of experiences and resources, the documentation of success stories, and the shaping of responses that facilitate recovery.

---

[1] Fort Myers is a city in southwestern Florida that lines the southern banks of the Caloosahatchee River.
[2] Papua New Guinea encompasses the eastern half of New Guinea and its offshore islands in the southwestern Pacific.

Decision-aid systems that improve coastal resilience include historical data, sensor networks, imagery, and models to provide people with actionable information. Table 2.3 provides an example of systems that are particularly useful to ocean and coastal engineers. Each relies on computers, communications, and data that allow scientists and engineers to build decision aid products to help solve marine-related problems. Data from these types of systems can be assimilated into numerical models or ingested into GIS. Table 2.4 highlights some of the state-of-the-art instrumentation and mobile data collection capabilities that allow for the collection of quickly perishable data, especially during post-disaster settings. Vessels may use these data to support the survey of damages following a maritime mishap (see Fig. 2.1). Data regarding impacts, responses by first responders, and environmental conditions following a disaster fill crucial gaps in engineering studies on how to be resilient. These sensor networks improve our ability to forecast hazards, make decisions in the face of uncertainty, enhance community resilience, and mitigate risk. They provide ground or sea truth data that are collected at scale for model skill assessments or to build realistic resilience planning scenarios.

2.2 Types of Hazards

Natural hazards include severe and extreme weather and climate events. In some cases, these hazards have been correlated with trends of increased surface, air, and ocean temperatures. For example, observational data in Fig. 2.2 indicates that the global mean sea level (GMSL) has risen by approximately 3.5 mm/yr since 1993. However, numerical modeling results (IPCC, 2013) indicated that the rate of rise could increase to somewhere between 8 mm/yr and 16 mm/yr before the end of the century depending on the extent of reductions in greenhouse gas emissions. Organizations such as NOAA have reported that the two major causes of GMSL are thermal expansion (water expanding from rising ocean temperature) and melting land ice (ice sheets and glaciers), which add water to ocean basins. Coastal ocean sea levels trends can diverge from GMSL in response to influence by major ocean currents such as the Gulf Stream and climate driven variations in the Atlantic Meridional Overturning Circulation (AMOC) system (Ezer, 2019). In recent decades coastal sea levels have been observed to be rising more rapidly than the global trend resulting in increased flooding hazards in some coastal areas (Ezer, 2020). For example, coastal ocean sea levels have been rising at rates of 10 mm per year over the past two decades along the east coast of Florida and elsewhere along the southeast U.S. coast (Dangendorf et al. 2023; Zarillo, 2023). This compares to longer-term rates of coastal sea level rise between 2 and 3 mm per year as recorded at NOAA water level stations. Within coastal relative sea level records are the local influences of land isostatic rebound or subsidence, sediment compaction, water, oil, and gas extraction, and land management contributing to relative changes in sea level. Isostatic adjustments are usually on the order of 1 mm per year or less, but the other factors can reach 1 cm per year or more.

Table 2.3 Selected sensor networks that provide data supporting decision makers

Operational network	Responsible agency
AErosol RObotic NETwork (AERONET)	NASA and Centre National d'Etudes Spatiales (CNES)
ARGOS Data collection and location system	NOAA and CNES
Automated Surface Observing Systems (ASOS)	NOAA/NWS
Australian Ocean Data Network (AODN)	Government of Australia
Clean Air Status and Trends Network (CASTNET)	EPA
Coastal Marine Automated Network (C-MAN)	NWS
European Global Ocean Observing System (EuroGOOS)	EU
Global Seismic Network (GSN)	U.S. Geological Survey (USGS), NSF, and EarthScope
Long term ecological research network	NSF
National Core Network (NCore)	EPA
National Mesonet	NWS
National streamflow network	USGS
National Water Level Observation Network (NWLON)	NOAA/NOS
NDBC ocean observation system	NOAA/NWS
Northwest Pacific Tsunami advisory center	Japan Meteorological Agency
Pacific Tsunami warning center	NOAA/NWS
Physical Oceanographic Real-Time Systems(PORTS)	NOAA/NOS
South China Sea Tsunami Advisory Center (SCSTAC)	People's Republic of China
Surge, Wave, and Tide Hydrodynamics (SWaTH) network	USGS

Natural hazards occurring along the coast become disasters when people's lives and livelihoods are destroyed. Numerous researchers (Herring et al., 2014; Manoj et al., 2020; Veeramony et al., 2020) have reported that increasing global surface temperatures will increase droughts and the intensity of storms. In the case of storms, the rise in atmospheric water vapor from evaporation fuels storms such as Typhoons Haiyan[3] (2013)

---

[3] Super typhoon Haiyan struck the Philippines on November 8, 2013.

Table 2.4 Key sensors that are used in operational networks to provide decision-aid information

Example sensors	Parameters	Example platforms
Acoustic, microwave, and pressure gages	Air gap, depth, water levels, waves	Bridges, piers, tide house, vessels
Accelerometers (wave buoys)	Wave heights, periods, and directions	Data buoys
Altimeters	The distance of a point above sea level	Aircraft, satellites
Anemometers	Wind direction and speed	Vessels, weather station
Barometer	Atmospheric pressure	Vessels, weather station
CTD	Conductivity, Temperature, Depth (variation of water temperature, salinity, and density)	Fixed (moored) and moving platforms (vessels)
Current meters (acoustic, electromagnetic, radar)	Current direction and speed	Data buoys, moorings, offshore platforms, vessels
DO sensors	Dissolved Oxygen	Data buoys, moorings, vessels
Echosounders, SONAR	Bathymetry, seafloor maps	Vessels
Fluorometer	Fluorescence	Data buoys, moorings
Hygrometer	Humidity	Weather station
Lidar	Shallow-water bathymetry	Stanchions, aircraft
Nephelometer	Turbidity and visibility	Data buoys, weather stations
Photometer	Sunlight and UV	Buoys, roof tops, towers
Radar	Surface currents	Bridges, vessels
Radiometers	Radiant (spectral) energy; sea surface temperature; ocean color	Aircraft, buoys, satellites, vessels
Radio	Lightning	Weather stations
Rain gauge	Rainfall totals	Weather stations
Salinometer	Salinity	Data buoys, piers, vessels
Thermistors	Temperature and humidity	Data buoys, moorings, vessels, weather stations,
Transmissometer	Visual range	Data buoys, moorings, stanchions

Vessels include ships, workboats, and autonomous platforms (e.g., gliders such as the Slocum glider; Unmanned Surface Vehicles such as Saildrone; and Autonomous Underwater Vehicles such as REMUS)

Fig. 2.1 U.S. Coast Guard Station Curtis Bay response crews preparing to deploy after the collapse of the Francis Scott Key Bridge in Baltimore, Maryland on March 26, 2024. Emergency response to this tragic disaster included local, state and federal agencies (Key Bridge Response 2024 Unified Command photo by U.S. Coast Guard Petty Officer 3rd Class Carmen Caver). NOAA deployed their Currents Real-time Buoys (aka CURBY) to support response operations in the Patapsco River, a tributary of the Chesapeake Bay

and Meranti[4] (2016) that formed in the Pacific Ocean. Tropical cyclones pose significant global threats to life and property from phenomena such as storm surges, flooding, extreme winds, and tornadoes. Examples of disasters are detailed in the International Disaster Database (EM-DAT) which contains data on the occurrence and impacts of over 26,000 mass disasters worldwide from 1900 to the present day (see https://www.emd at.be/). NOAA's National Centers for Environmental Information maintains a map service called Global Natural Hazards Data (see https://hub.arcgis.com/maps/b146357d106e 4cbfa9e9c41fd0f362b3/explore) for earthquakes, tsunamis, and volcanoes. These types of databases are essential for researchers and help disaster responders improve their planning (mitigation, preparedness, response, and recovery) and emergency management.

Natural hazards span a range of temporal and spatial scales, a concept which is especially important to those concerned about risk and vulnerability. One might consider a heatwave that crosses a particular temperature threshold for a city, on a day or a few

---

[4] Super Typhoon Meranti impacted the northern Philippines, Taiwan and China during September 2016.

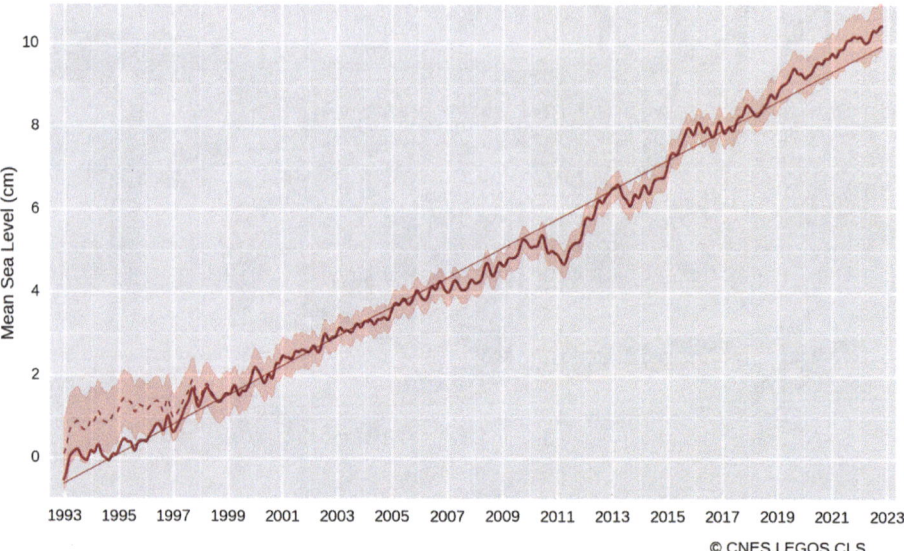

Fig. 2.2 Observed sea level rise curve based on satellite altimetry (Ablain et al., 2019). Data were collected from the TopEx/Poseidon, Jason-1, Jason-2 and Jason-3 missions from January 1993 to present. Analysts applied a postglacial rebound correction (–0.3 mm/yr) to estimate a rise in mean sea level of 3.6 mm/year with an uncertainty of 0.4 mm/yr

days, rather than estimating country scale heat extremes. Multidisciplinary and integrated approaches are needed to explain the spatial–temporal patterns of natural hazards, processes and mechanisms, emergency responses, risks, and mitigation of natural hazards. For example, scientists might use satellite altimeters to measure absolute sea level change referenced to the height of the ocean surface above the center of the earth and tide gages to record relative sea level change near a particular location. Tide gages are especially important to measure high tides and recurring floods that may occur during spring and King tides. Spring tides are the high tides that result from a new or full moon. King tides are actually perigean spring tides, which only occur when the moon is closest to the Earth (perigee). The ensuing nuisance floods are a precursor to future flooding from sea level rise and provide valuable information for land use planning, coastal engineering, emergency preparedness, and community behavior. Long-term data are needed to understand the frequency, duration, extent, and speed of onset for a particular hazard and the more complex situation when the hazard combines with other hazards (e.g., droughts and wildfires; earthquakes and landslides; and tropical cyclones and floods).

Although hazards occur in all parts of the world, some coastal regions are more vulnerable to hazards than others. Deltas, which were described by Hart and Coleman (2005) in the World Delta Database (see https://www.geol.lsu.edu/WDD/), are at risk from multiple threats. Since these deltas form at or below sea level, storm surges are expected to occur more frequently. They are especially susceptible to land subsidence and relative sea-level rise and tend to be associated with high population densities. For example, the Bay of Bengal, situated on the Brahmaputra River Delta, is prone to flooding. Warrick et al. (1996) pointed out that the increased frequency and intensity of cyclones and storm surges from climate change would amplify the threat of coastal flooding in the Bay of Bengal. Cyclones, floods and monsoons can cause destruction to low-lying villages and where wetlands have been converted to agricultural use in countries such as Sri Lanka and India to the west, Bangladesh to the north, and Myanmar to the east.

Measuring relative sea level rise is most important and includes processes such as subsidence. Vinogradova and Hamlington (2022) described benefits of observations by NASA and partner organizations that illustrate how fast sea level is changing near coastal cities around the world. The City of New Orleans[5] provides a good example since relative sea level is rising very rapidly with respect to the land, which is sinking at up to 18 mm/ yr. in parts of coastal Louisiana (Allison et al., 2016). For deltas and wetlands, relative sea level rise of 5 mm is considered to be the "tipping Point" for transitioning from vegetated wetlands to open water bays (Törnqvist et al., 2020). Erosion and shoreline retreat will continue and are attributed to rises in sea levels from a warmer global climate (Hauer et al., 2021). Determining rates of shoreline retreat provides valuable information for land use planning, coastal engineering, emergency preparedness, and community behavior. Owing to population concentrations along the coast, understanding relative sea level rise and the resulting rates of shoreline retreat are important factors for local governments, scientists, engineers, and citizens.

Important lessons from Hurricane Ian (2022) and other recent storms have been learned through effective collaboration among scientists, engineers, and the community. We have learned that storm intensification is becoming more rapid, as storms during the last decade have grown from tropical storms to major hurricanes over periods of one or two days (e.g., Klotzbach et al., 2022). This considerably shortens the lead times of forecasts and the preparation for extreme weather. In addition, the storms are becoming larger and are increasingly accompanied by torrential rainfall, which adds to storm surges to produce compound flooding (Pietrafesa et al., 2019). To date, there is no significant evidence that tropical cyclones are becoming more frequent, just that they develop faster and become larger and stronger than in the past (see Sect. 3.1). Observing systems such as GCOOS and SECOORA, which are a part of the Global Ocean Observing System provide evidence that the storms that do form and make landfall are more dangerous to coastal cities such as Fort Myers, which was devastated by Hurricane Ian. Hurricane-related information can

---

[5] New Orleans in Louisiana is near the Gulf Mexico. The city sits between the Mississippi River to the south and Lake Pontchartrain to the north.

Table 2.5 Ocean observatories that are sponsored by the NOAA U.S. IOOS program office. They are being operated as the US' component of the global ocean observing system

Observatory	URL
Alaska ocean observing system	https://ioos.noaa.gov/regions/aoos/
Caribbean coastal ocean observing system	https://ioos.noaa.gov/regions/caricoos/
Central and Northern California ocean observing system	https://ioos.noaa.gov/regions/cencoos/
Gulf of Mexico coastal ocean observing system	https://ioos.noaa.gov/regions/gcoos/
Great lakes observing system	https://ioos.noaa.gov/regions/glos/
Mid-Atlantic regional association coastal ocean observing system	https://ioos.noaa.gov/regions/maracoos/
Northwest association of networked ocean observing systems	https://ioos.noaa.gov/regions/nanoos/
Northeastern regional association of coastal ocean observing systems	https://ioos.noaa.gov/regions/neracoos/
Pacific islands ocean observing system	https://ioos.noaa.gov/regions/pacioos/
Southeast Coastal ocean observing regional association	https://ioos.noaa.gov/regions/secoora/
Southern California coastal ocean observing system	https://ioos.noaa.gov/regions/sccoos/

These regional observatories include stakeholders from universities, government, and some marine science and technology companies. They are strategically located to serve coastal communities in the United States, including the Great Lakes, the Caribbean and the Pacific Islands and territories

be accessed online from GCOOS' hurricane information dashboard (see https://geo.gcoos. org/hurricane/) or SECOORA's hurricane resources (see https://secoora.org/hurricane-res ources/). Eleven observatories that comprise the U.S. IOOS are listed in Table 2.5. They provide the infrastructure to track, predict, manage, and adapt to changes in our coastal environments.

2.3   Coastal Examples

The coastal zone is the interface between the land and water. These areas are formed by fundamental physical and biological processes such as sea ice, winds, waves, climate, oyster reefs, mangroves, corals, and more. Coasts are also significant because most of the world's population inhabit these areas. Infrastructure includes airfields, bridges, platforms, towers, seaports, communities, and cities. Wright and Nichols (2019) highlighted that coasts are continually changing owing to their dynamic interaction with oceanic and land processes and the long-term impact of climatic changes such as sea level rise. Energy reaching the coast can become high, especially during storms, and high energy events make the coastal zone an area of high vulnerability. Some example disasters from natural hazards are provided in Table 2.6.

Climate change has already influenced the frequency and severity of dangerous fire conditions around the world (Abram et al., 2021; Carnicer et al., 2022; Goss et al., 2020). Fires have been shown to impact estuarine health, for example in New South

Table 2.6 Example natural hazards and ensuing disasters

Phenomena	Example
Dead zones	Dead zones occur in coastal regions owing to nutrient pollution. The development of hypoxic conditions or "dead zones" in the northern Indian Ocean was described by Naqvi and Wajih (2021). Extensive dead zones have been mapped in the Arabian Sea and Gulf of Mexico
Debris flows	A catastrophic debris flow on January 9, 2018, in Montecito, California killed 23 residents and caused massive damage (Burns, 2022)
Drought	The 2009–2010, 2014–2016, and 2019–2020 drought events in the Caribbean reduced the amount of freshwater entering estuaries and increased salinity levels, which altered habitat conditions and led to declines in the richness and abundance of fresh and brackish water species (Trotman et al., 2018; United Nations Office for Disaster Risk Reduction, 2021)
Extreme waves	Rogue waves, which are more than twice the height of the surrounding waves, have been shown to be responsible for the sinking of vessels (Gemmrich & Cicon, 2022). Rogue waves have been reported by ships such as the British ocean liner *Queen Elizabeth II* on September 11, 1995, measured by instrumented buoys, and simulated by numerical models
Forest fires	Canadian wildland fires during Summer 2023 reduced visibility and air quality across vast sections of North America. Sidik (2020) associated wildfire severity and frequency in Canada to climate change
Floods	Mediterranean Storm Daniel passed through eastern Libya over the weekend of Sept. 9, 2023, bringing heavy rainfall and flooding that resulted in large-scale destruction (WMO, 2023)
Harmful algal blooms	A bloom of marine plankton (*Pseudo-nitzschia*) during 2015 along the U.S. West Coast caused widespread mortality of marine mammals and seabirds and contamination of shellfish (Ekstrom et al., 2020)
Heatwaves	Heatwaves, which are well documented in Australia, pose a significant hazard to people and the environment. Patel (2022) described extreme heat events occurring during January 2022 and Nairn et al. (2021) described Australia's Black Summer heatwaves
Tropical cyclones	Tropical Cyclone Bhola struck Bangladesh from 12–13 November 1970 and killed more than 300,000 people (Longshore, 2007; WMO, 2020)

Note The occurrence of natural hazards (e.g., floods and tropical cyclones) which cause damage or loss of life has increased during the past several decades (e.g., Coronese et al., 2019; IFRC, 2020; Institute for Economics & Peace, 2020; Petryk, 2023)

Wales,[6] Australia (Barros et al., 2022). Particulate matter from fires in coastal regions alter water chemistry and increase silt content. In addition to draught conditions causing longer and more severe fire seasons during 2022, the concomitant low water conditions in European rivers challenged navigation. Reduced mountaintop snowpack and increased lightning strikes are two other reasons associated with climate change, which contribute to dangerous fire conditions. The development of viable fire management programs is a concern for government, companies, communities, and citizens who have an interest in or stake in local resources and resilient outcomes.

Engineers, environmental scientists, and communities must work together to mitigate the impact of dangerous mixtures of sediment and water that might flow down a slope under the influence of gravity. Such debris flows or landslides may start along elevated areas and generally follow periods of intense rainfall or rapid snowmelt along some sort of favorable drainage area (Mueting et al., 2021). Stock and Dietrich (2006) described debris flows along the Oregon coast occurring in response to deforestation and rainfall. Down slope areas that have been deforested (e.g., burned by a forest fire) are especially susceptible to debris flows. According to Lin et al. (2022), improved resilience to debris flows requires consideration of the physical environment, society and economy, system and facilities, and adjustment and learning. Numerical models can accurately represent the physics of debris-flow and these kinds of simulations can be used to implement early warning systems.

Climate change alters the timing of water availability. Longer periods of drought and warmer temperatures that enhance evaporation result in decreased water levels and flows in water bodies such as streams, lakes, and rivers. Severe droughts during Summer 2022 reduced water levels in Lake Mead[7] and altered navigation patterns in rivers such as the Rhône River in France (Toreti et al., 2022). In addition to transportation along major rivers, these droughts impact agriculture and energy production. Extended periods of drought threaten the water supply and impact the health, safety, and welfare of communities. Severe drought conditions are currently contributing to food shortages in the Horn of Africa (Djibouti, Eritrea, Ethiopia, Kenya, Somalia, South Sudan, Sudan, and Uganda). Researchers such as Barthel and Seidl (2017) have highlighted collaborative efforts between physical and social scientists, which have been attributed to the solution of complex groundwater problems. Cooperation among government, university, industry, and community stakeholders promotes the sustainable use of this resource.

Extreme waves may be caused by major storms at sea or the interaction of waves with currents. Most extreme waves are caused by superposition or the combination of two waves at the same location, where constructive interference involves two identical waves being superimposed in phase and destructive interference involves two identical waves

---

[6] New South Wales bordering the Coral and Tasman Seas has well-defined seasons that feature hot summers and cool winters and spring and autumn transitions.

[7] Lake Mead in Arizona and Nevada is a reservoir formed by the Hoover Dam on the Colorado River (Hannoun & Tietjen, 2022).

being superimposed exactly out of phase. Since constructive interference leading to rogue waves is uncommon, the measurements of this phenomenon are rare. Rogue waves that reach shore are called sneaker waves[8] in regions such as the Pacific Northwest. Exactly how and when rogue waves form is still under investigation and numerical modeling has provided advances (Manzetti, 2018). Fanti et al. (2023) described a tendency for underestimating extreme wave heights in wave hindcasts and highlighted the importance of robust wave buoy networks. The accurate prediction of extreme waves is vital for issuing warnings, protecting property from erosion, and saving lives. Since extreme waves are especially dangerous to ships and isolated structures, improved knowledge on the occurrence of extreme waves is particularly important to mariners, ocean engineers, naval architects, and those involved in optimal ship routing.

Floods are common natural disasters caused by too much water in a location. They are generally caused by prolonged rainfall, rapid melting of large amounts of snow or ice, storm surge, or the bursting of dams or levees. Risks from flooding increase as populations continue to grow and urban areas expand. Lee et al. (2020) highlighted that from 2001 to 2018, approximately 600 billion USD were lost globally due to over 2,900 floods and that nearly 300,000 people suffered from flood-related physical injuries during this period. Resilience requires understanding the different kinds of floods, which is key to forecasting their occurrence and mitigating the risk. Traditional structures used to manage floods include dams, retention basins, levees, seawalls, and weirs. First responders require special training to deal with flood impacts such as swift water rescue, which involves rescuing people from fast-moving water. While flooding is often disastrous, it can be managed through structural and nonstructural approaches.

Global warming allows the atmosphere to hold more water while urbanization contributes to more runoff. These facts contribute to heavier precipitation and an increased likelihood of flash flooding. In addition to excessive rainfall, heavy snow melts may also cause rivers to flood as their capacity is exceeded. Fluvial or river floods can cause widespread damage, e.g., when dams, dikes, or levees break. Such floods are characterized by an intense, high velocity torrent of water that generally occurs in an existing river channel with little to no notice. Fast moving floods are very dangerous and destructive not only because of the force of the water, but also hurtling debris such as cars that are often swept up in the flow. Flooding can also cause major problems with septic systems[9] owing to soil saturation. Failing septic systems can discharge untreated wastewater onto the ground and into surface waters (Herren et al., 2021; Withers et al., 2014). Water reservoirs including the oceans, lakes, rivers, streams, glaciers, and snowfields are inextricably linked to climate change.

---

[8] Garcia and Castleman (2023) provided a short discussion and videos of sneaker waves occurring along some California beaches.

[9] A septic system includes an underground concrete, fiberglass, or plastic holding tank, pipes, and a drainage field for domestic sewage treatment.

A pluvial, or surface water flood, is caused when heavy rainfall creates a flood event independent of an overflowing water body. Tonn and Czajkowski (2022) discussed the risk and complexity of pluvial flood damage in the United States. They described pluvial flooding in Harris County Texas[10] as the result of factors which included Hurricane Harvey (2017) rainfall, flat topography, soils, and stormwater management practices. The intense rain saturated Houston's drainage systems. As the system became overwhelmed, water flowed out into streets and nearby structures. For these reasons, city planners should work with expert partners, such as national or regional meteorological departments, universities and other research institutions, as well as relevant NGOs to develop flood maps and risk assessments that ensure that the city will be able to adapt to future climate change scenarios.

Low-lying built-up areas such as Ellicott City in Maryland are prone to floods. The Patapsco River, which flows through Ellicott City (a city that was founded in 1772), is the primary culprit along with development along the watershed. Significant flooding occurred during 1868, 1901, 1917, 1923, 1942, 1952, 1972, 2016 and 2018. Flooding during 1972 occurred when the Patapsco River overflowed onto the neighboring land owing to excessive rain and storm surge from hurricane Agnes. Deadly flash floods occurred in response to torrential rainfall on July 30, 2016, and May 27, 2018. Efforts to mitigate the flood threat included the installation of a detention pond during 2023 that can divert floodwaters away from Ellicott City's Main Street and hold up to 3.3 million gallons of water during storm events.

Coastal flooding occurs when water inundates coastal lands as a result of extreme high or rising tides, storm surges, or the right combination of both. Coastal flooding during Hurricane Katrina (2005) and Sandy (2012) have been well studied (Lander, 2022; Plough & Chandra, 2015). The failure of levees containing the Mississippi River Gulf Outlet[11] was the reason so much water flooded into the 9th Ward of New Orleans during Katrina. The severe impact was much worse than during Camille in 1969 even though Camille was Category 5 and Katrina Category 3 at landfall because of the loss of protective wetlands between 1969 and 2005. Sandy destroyed shore protection, caused mudslides, filled subways, and destroyed more than 69,000 residential units in New York. Strauss et al. (2021) quantify Sandy damages in terms of past storms and differing rates of sea level rise. In the United States, the Federal Emergency Management Agency (FEMA) has developed a Coastal Flooding Risk Index which represents a community's relative risk to coastal flooding (see https://hazards.fema.gov/nri/coastal-flooding). Some related and accessible online resources that support the development of risk assessments are provided in Table 2.7. Organizations such as the United Nations Educational, Scientific and Cultural Organization (UNESCO) facilitate international cooperation on resilient solutions to natural hazards such as flooding, worldwide.

---

[10] Houston is a large flood-prone city in Texas, which is located primarily in Harris County.

[11] The Mississippi River–Gulf Outlet is a 76 mi canal constructed by the USACE (see https://www.mvn.usace.army.mil/Missions/Environmental/MRGO-Ecosystem-Restoration/History-of-MRGO/).

Table 2.7 Accessible resources for scientific data and coastal-related risk assessments

Resource	Responsible organization	URL
Climate central	Climate central	https://www.climatecentral.org/resources?tab=content
Climate risk & resilience portal	Argonne National Laboratory	https://www.anl.gov/ccrds/ClimRR
Coastal flood exposure mapper	NOAA	https://coast.noaa.gov/digitalcoast/tools/flood-exposure.html
European marine observation and data network	European commission	https://emodnet.ec.europa.eu/en
National risk map	FEMA	https://hazards.fema.gov/nri/map
Risk factor	Risk factor™	https://help.riskfactor.com/hc/en-us
Sea level analysis tool	USACE	https://climate.sec.usace.army.mil/slat/

Wildfires depend on factors such as air temperature, low humidity, soil moisture, and flammable vegetation. They originate from unplanned ignition, such as lightning, volcanos, and unauthorized and accidental human caused fires (NWCG, 2020). Climate variations that enhance the drying of organic matter in forests have been attributed to an increasing number of wildfires (Moore, 2022; Stevens-Rumann et al., 2018; Westerling et al., 2006). Canada had a particularly high level of wildfire activity during 2023 following the development of large fires in Alberta, British Columbia, and Saskatchewan (Bassler, 2024). While eliminating the risk of future wildfires is not feasible, integrated approaches that promote sustainable land use and discourage deforestation can mitigate some of the wildfire drivers. Organizations such as the National Center for Atmospheric Research are collaborating with researchers worldwide to explore how a warming climate affects the incidence of wildfires and impacts air quality (e.g., Fasullo et al., 2021). Wildfire management should include elements related to risk mitigation and prevention, fire suppression, and post-disaster recovery plans.

Harmful Algal Blooms (HABs) occur when colonies of algae known as phytoplankton grow out of control and produce toxic or harmful effects on fish, shellfish, birds, marine mammals, and people. There are many types of blooms, and some are responsible for phenomena known as brown tides (*Aureococcus anophagefferens*), red tides (dinoflagellates, Fig. 2.3), and white tides (coccolithophores). The blooms will drift around due to tides, winds and currents. HABs have gained a lot of attention due to their impact on water quality and debilitating or even fatal illnesses that people can acquire through contact.

Fig. 2.3 NOAA photograph of a "red" tide off the coast of Texas. Organizations such as NOAA issues forecasts to help people make informed choices about where and when to visit areas that may be temporarily affected by a HAB (see https://oceanservice.noaa.gov/hazards/hab/gulf-mexico.html)

Barker (1997) described outbreaks of *Pfiesteria piscicida* in North Carolina[12] in his book, *And the waters turned to blood*. Another example of HABs are formed by cyanobacteria (blue-green algae), which thrive in warm, slow-moving water. Consequently, warming waters in response to climate change are expected to increase the magnitude and duration of cyanobacteria blooms. Such blooms are a worldwide phenomenon as evidenced by research with implications for the Baltic Sea (Munkes et al., 2021), Lake Okeechobee in Florida (e.g., Rosen et al., 2017), and Lake Taihu in China (e.g., Qin et al., 2021). HABs tend to cause dissolved oxygen swings[13] that may result in plant and animal die-off and public health issues. Some algal blooms will wash up onshore and decompose, which provides a potential threat to beach goers. If blooms are detected, people should be warned to stay out of the water as well as their pets and livestock. Health officials will issue alerts and close commercial shellfish beds owing to threats such as paralytic shellfish poisoning (Ethridge, 2010), which is attributed to saxitoxin, a neurotoxin naturally produced by certain species of dinoflagellates and cyanobacteria.

Heat waves occur when high pressure aloft strengthens and remains over a region for several days up to several weeks. The elevated temperatures and lack of rainfall cause problems ranging from water shortages and increased stress on plants to heat exhaustion and heat stroke for people. Extreme heatwaves may exceed the temperature thresholds of materials and equipment that comprise structures. According to Ballester et al. (2023), the heatwave of June to August 2022 in Europe showed that society is struggling with climate changes as evidenced by about 63,000 heat-related deaths. Other impacts included

---

[12] North Carolina is a cuspate foreland that includes barrier islands interrupted by coastal inlets, and the Albemarle and Pamlico Sounds (see https://coast.noaa.gov/nerrs/reserves/north-carolina.html).

[13] Organic and nutrient enrichment related to sewage/industrial discharges and land runoff contributes to hypoxia and "dead zones" that form along vast stretches of coastal waters (Diaz & Rosenberg, 2011).

low river discharge. Major rivers such as the Rhine and Danube had low water levels, which prevented some vessels from navigating the waterways. Impacts on fisheries such as Pacific sardines are reported by a high-pressure system called "the Blob" in the North Pacific (Free et al., 2023). Barkhordarian et al. (2022) attributed the phenomena to the Pacific Warm Pool[14] and global warming. In some counties heat waves and their consequences have contributed to the development of heat prevention plans and other adaptation strategies aimed at protecting vulnerable populations.

Tropical cyclones form over large, warm bodies of water and can produce torrential flooding, damaging winds, tornadic activity, erosion, hail, storm surge, and coastal flooding. Tropical cyclone is the generic term for a hurricane which forms in the Atlantic or northeastern Pacific Ocean, a typhoon which occurs in the northwestern Pacific Ocean, or a cyclone which occurs in the Indian Ocean and South Pacific. These storms carry heat and energy away from the tropics and transport it towards temperate latitudes, which plays an important role in regulating climate. Tropical cyclones are typically between 100 and 2,000 km in diameter. While there are many ways to characterize a tropical cyclone, Typhoon Tip (1979), with a central pressure of 870 mbar and a diameter of 2,220 km, is the largest tropical cyclone on record. This typhoon generated peak wind speeds of 300 km/hr on October 12, 1979 (Evans & AccuWeather, 2012). This typhoon sunk ships and made landfall in Japan where it caused mudslides, destroyed structures such as bridges and dikes, and flooded homes (Iacovelli & Vasquez, 1998). Efforts to improve resilience requires collaboration on efforts such as pre-disaster planning and hazard mitigation; building storm-resistant structures; improving forecasts and communications; and supporting managed retreat from highly vulnerable coastal areas. Programs such as the Rockefeller Foundation's 100 Resilient Cities (see https://www.rockefellerfoundation.org/100-resilient-cities/) have been instrumental in facilitating effective collaboration among government, university, and industry partners.

2.4 Effects of Climate Change

Climate change refers to long-term shifts in temperatures and weather patterns. Climate variability is nothing new but needs to be handled responsibly. The general effects are detailed in Table 2.8 with recent examples. Warming temperatures over time are changing hardiness zones for agriculture and even the utility of traditional construction materials. With respect to structures and buildings, expansion and contraction of materials can lead to the development of cracks, gaps, and weaknesses in a structure over time. Engineers must manage temperatures and resilient applications involve the effective recovery and reuse of heat that would otherwise be wasted. Mitigating some of these effects would require collaboration, where engineers would engage with environmental scientists and

---

[14] The waters of the Pacific Warm Pool are warmer than any other open ocean on Earth (De Deckker, 2016).

the community to increase shade around buildings, use green or cool roofs in construction, and install energy-efficient appliances and equipment. Yan et al. (2024) provided a review of green roof technology. At a societal level, climate change requires governance to identify risks and ensure a viable communication system that facilitates the communication of alerts and warnings. Climate change has the tendency to create or exacerbate social injustices as illustrated by low income or impoverished people living in areas subjected to effects such as droughts or flooding.

A critical aspect in the design of resilient coastal and oceanic structures involves the estimation of environmental conditions. Nichols and Raghukumar (2020) discussed the use of long-term ocean observatories, unmanned platforms, satellite and coastal remote sensing imagery, data assimilative numerical models, and high-speed communications to provide operators with actionable environmental information. Special attention must be

Table 2.8 Selected phenomena attributed to climate change, consequences, recent examples, and evidence

Phenomenon	Consequences	Evidence and examples
Food scarcity	Social and political conflicts	Sova and Zembilci (2023) described the feedback loop between conflict and hunger that drives war
Hotter temperatures	Changing rainfall patterns, more destructive storms, sea level rise, heat exhaustion	Average surface air temperature increases are documented by weather stations, data buoys, and ships and from infrared detection by satellites (Susskind et al., 2019)
Human displacement	Abandonment of coastal structures	The Internal Displacement Monitoring Center (n.d.) reported that 8.7 million people were living in internal displacement due to disasters at the at the end of 2022
Increased drought	Habitat loss, drinking water shortages, food insecurity	Rodell and Li (2023) characterized prolonged extreme droughts by measuring groundwater changes from space
More health risks	Foodborne illnesses, waterborne diseases, mosquito-borne diseases	The EPA (2024) provided an explanation on how climate change impacts human health and well-being through the spread of pests and diseases
More severe storms	Stronger winds, heavier rainfall, increased flooding	There's evidence (Zhu & Quiring, 2022) that global warming has already been increasing the amount of rain from storms such as Hurricane Harvey (2017)
Sea level rise	Shoreline retreat, marsh migration	Lindsey (2022) reported that global mean sea level has risen about 8–9 inches (21–24 cm) since 1880 owing to a combination of melt water from glaciers and ice sheets and thermal expansion of seawater as it warms

Additional information can be obtained from resources such as the United Nations (see https://www.un.org/en/climatechange)

taken to using these data to determine environmental loads from natural events that can cause damage such as earthquakes, floods, and tropical cyclones, which are all difficult to predict or quantify for design. Flood waters are known to damage roads by eroding underlying unbound granular layers, washing out road sections, damaging the bituminous layers, reducing load bearing capacity, reducing the support in concrete pavements, etc. Engineers working with government and NGOs can protect important coastal roads by implementing a range of resilient efforts from increasing roadway elevation to creating living shorelines to installing sensors that warn motorists of potential flooding events. The Federal Highway Administration (2016) described the use of living shorelines in Brookhaven, New York[15] as an engineering case study. The case study pointed out that there are approximately 60,000 miles of road in the United States that are occasionally exposed to coastal storm surge and waves. Coastal flooding along these transportation routes can also affect human health by increasing the risk that drinking water and wastewater infrastructure will fail, putting people at risk of being exposed to contaminated water.

There are many weather-related structural failures that are caused by the effects of climate change—high temperatures, rainfall, icing, storms, and lightning. During August 2023, wildfires that broke out on the island of Maui in Hawaii were most likely caused by downed power lines, which ignited tall grass and brush and spread owing to favorable wind conditions (Boesecker et al., 2023). The wildfires were attributed to meteorological drivers such as dry, gusty conditions created by a strong high-pressure area north of Hawaii and Hurricane Dora to the south (Julaino et al., 2024). While natural causes such as lightning and volcanic activity might cause wildfires, most are caused by humans. Catastrophes such as wildfires, which are linked to extreme weather conditions, are an impact of climate change. These kinds of risks must be considered seriously by engineers, environmental scientists, and communities. To prevent mishaps and structural failures, engineers must consider both static and dynamic forces acting on structures. It is particularly important for engineers to consider extreme dynamic forces produced by the environment, e.g., wind and waves, which are responsible for many structural failures.

Future climatic conditions will impact coastal infrastructure that includes bridges, towers, wind turbines, offshore platforms, seafloor cables, and moored data buoys. More frequent tropical cyclones, wetter winters, and heavy downpours are example phenomena that increase the risk of damage. The design, impact, and maintenance of structures that will be installed in coastal waters that are subjected to highly varying environmental loads ranging from strong currents to breaking waves to ice flows must be carefully considered among multiple interest groups. The impacts of extreme weather on man-made objects (e.g., wave-slamming forces acting on piers and bridge decks, both vertical and inclined piles, and even along the shoreline) have consequences for the environment, travelers, and the community. Ongoing threats such as erosion (the general lowering of the ground

---

[15] Brookhaven, New York is a town on Long Island that touches both the Atlantic Ocean and Long Island Sound.

surface over a wide area), which may be caused by wave action and currents may require the implementation of very expensive beach nourishment programs. Further, scour which refers to a localized loss of soil, often around a foundation element such as pilings, may require the procurement of armoring stone to safeguard critical infrastructure such as bridges.

As waves are breaking higher on the beach, shoreline retreat results during periods when strong undertow currents from wave breaking evacuate beach sand to offshore areas. The rising water levels inundate lowlands, displace wetlands, and alter the tidal range in rivers and bays. Arns et al. (2017) highlighted that shore protection strategies must reconsider the design heights of shore protection owing to factors such as sea level rise. An additional effect that is monitored very closely in Florida is the increased salinity that can result from saltwater intrusion into aquifers and estuarine systems. Shaw and Zamorano (2020) described the extent of seawater encroachment into aquifers along the South Florida coastline. Thus, the impacts of climate-induced sea level rise are likely to worsen problems such as coastal flooding, shoreline erosion, and water quality, which will require the development of innovative coastal protection.

Climate change is causing an increase in the frequency and severity of natural disasters, such as tropical cyclones, wildfires, and droughts. Newman and Noy (2023) evaluated 185 events in the International Disaster Database (see https://www.emdat.be/) and estimated a net cost of $260.8 billion in economic damages owing to climate change during the period from 2000 to 2019. The increased frequency of extreme events can cause significant damage to properties, which can in turn disrupt real estate markets and quality of life. Losses and damage may permanently alter the affected region. For example, rising seas encroaching low-lying islands such as Smith Island in the lower Chesapeake Bay (Rehak, 2024) or the Maldives in the Indian Ocean (Jaschik, 2014) threaten whole communities. Saltwater intrusion threatens drinking water in places such as Florida and turns once-productive farmland in Venice, Italy into unproductive barren land (Tosi et al., 2022). Elevated salinity levels in the soil cause crop yield declines, coastal forest loss, and marsh migration.

Organizations such as the World Health Organization provide evidence that climate change causes issues ranging from the availability of food to vector-borne diseases such as Zika. According to the University of Notre Dame's Global Adaptation Initiative (ND-GAIN) Country Index, the Caribbean's food systems are vulnerable to climate change based on factors such as agricultural capacity and food import dependency (see https://gain.nd.edu/our-work/country-index/). Owing to the extent of coastal lands in the Caribbean, the greater number and frequency of extreme weather events, long droughts, landslides and floods, and coastal erosion impact agriculture. Van Wyk et al. (2023) described the highly temperature-dependent transmission risks of Zika and dengue fever in different climate change scenarios in Brazil. In addition to temperature, extreme rainfall events or flooding may produce standing water that harbors pollution, harmful bacteria, viruses, parasites, fungi, and the perfect breeding ground for mosquitoes. The health effects of climate change are many and have been directly linked to water- and food-related illnesses, injuries, and deaths.

Resilience requires engineers, scientists and the community to consider climate change factors such as sea level rise. Numerous authors have described sea level rise (Cooper & Pilkey, 2004; Inman & Dolan, 1989; Khojasteh et al., 2023), which poses a serious threat to coasts around the world (Melet et al., 2023; Wright & Nichols, 2019). Example consequences include increased intensity of storm surges, changes in locations of wave breaking along the coast, damaged changing flood patterns, inland movement of the salt waterfront, and damage to structures and the fragile wetland areas. In many cases, this is where large population centers are located, in addition to important estuarine habitats. Recurring floods of coastal roads and parking lots associated with high tides is reminder of the impact of sea level rise. Evidence of the patterns of sea level rise are observed by sensors from coastal weather stations and water level gage networks. These data suggest the elevation or movement of critical structures and the redesign of coastal protection. In other locations as illustrated in Fig. 2.4, property is already lost to the impact of inadequate structures and climate change.

Fig. 2.4 Landslide, Scarborough North Yorkshire that destroyed the Holbeck Hall hotel between the nights of 3 and 5 June 1993 (Pennington, n.d.). Factors contributing to the landslide included excessive rainfall and coastal erosion

Fig. 2.5 An abandoned home and flooding on Wingate-Bishops head road on the Eastern Shore of Maryland during an October 2007 spring high tide (Marra, 2015)

An effect that is seen throughout the world involves the abandonment of coastal properties that have been subject to severe flooding. Lincke and Hinkel (2021) described this effect based on sea-level rise scenarios at a global scale. In the United States many areas where one might find abandoned properties lie along the Gulf of Mexico coast (e.g., in Texas and Florida) and along the mid-Atlantic (e.g., abandoned properties located along the coastline from Montauk Point to Virginia Beach, which also includes shorelines associated with the Chesapeake Bay and the lower Delaware River). Paterson (2019) highlighted the threat for historic sites in Annapolis, Maryland and critical infrastructure such as the U.S. Naval Academy. Following inundation, there is likely damage to structures, many dwellings will be unlivable, and some houses will be condemned. There is often no uniform answer on liability for abandoned structures such as in Fig. 2.5. If owners are not held responsible for abandoned property, who should pay to safeguard that environment and remove potentially dangerous structures? In response to damages occurring after hurricanes such as Authur in 2014 (Fig. 2.6) and Matthew in 2016, North Carolina's Office of Recovery and Resiliency implemented a strategic buyout program called ReBuild NC (see https://www.rebuild.nc.gov/homeowners-and-landlords/strategic-buyout-program). The program offers fair-market-value buyouts to homeowners located in areas at high risk of damage from hurricanes and floods.

Fig. 2.6 North Carolina's outer banks barrier islands after the passage of Hurricane Arthur during July 2014 (U.S. Coast Guard photo by Petty Officer 3rd Class David Weydert)

References

Ablain, M., Meyssignac, B., Zawadzki, L., Jugier, R., Ribes, A., Spada, G., Benveniste, J., Cazenave, A., & Picot, N. (2019). Uncertainty in satellite estimates of global mean sea-level changes, trend and acceleration. *Earth System Science Data, 11*(3), 1189–1202. https://doi.org/10.5194/essd-11-1189-2019

Abram, N.J., Henley, B.J., Gupta, A. S., Lippmann, T. J. R., Clarke, H., Dowdy, A. J., Sharples, J. J., Nolan, R. H., Zhang, T., Wooster, M. J., Wurtzel, J. B., Meissner, K. J., Pitman, A. J., Ukkola, A. M., Murphy, B. P. Nigel, Tapper, J. & Boer, M. M. (2021). Connections of climate change and variability to large and extreme forest fires in southeast Australia. *Communications Earth & Environment, 2*, 8, https://doi.org/10.1038/s43247-020-00065-8

Allison, M., Yuill, B.,Törnqvist, T., Amelung, F.,Dixon, T. H., Erkens, G., Stuurman, R., Jones, C., Milne, G., Steckler, M., Syvitski, J., & Teatini, P. (2016). Global risks and research priorities for coastal subsidence. *Eos, 97*. https://doi.org/10.1029/2016EO055013

Arns, A., Dangendorf, S., Jensen, J., Talke, S., Bender, J., & Pattiaratchi, C. (2017). Sea-level rise induced amplification of coastal protection design heights. *Scientific Reports, 7*, 40171. https://doi.org/10.1038/srep40171

Ballester, J., Quijal-Zamorano, M., Méndez Turrubiates, R. F., Pefenaute, R., Herrmann, F. R., Robine, J. M., Basagaña, X., Tonne, C., Antó, J. K., & Achebak, H. (2023). Heat-related mortality in Europe during the summer of 2022. *Nature Medicine, 29*, 1857–1866. https://doi.org/10.1038/s41591-023-02419-z

Barkhordarian, A., Nielsen, D. M., & Baehr, J. (2022). Recent marine heatwaves in the North Pacific warming pool can be attributed to rising atmospheric levels of greenhouse gases. *Communications Earth & Environment, 3*(1), 1–12. https://doi.org/10.1038/s43247-022-00461-2

Barker, R. (1997). *And the waters turned to blood*. Simon & Schuster

Barros, T. L., Bracewell, S. A., Mayer-Pinto, M., Dafforn, K. A., Simpson, S. L., Farrell, M., & Johnston, E. L. (2022). Wildfires cause rapid changes to estuarine benthic habitat. *Environmental Pollution, 308*, 119571. https://doi.org/10.1016/j.envpol.2022.119571

Barthel, R., & Seidl, R. (2017). Interdisciplinary collaboration between natural and social sciences—Status and trends exemplified in groundwater research. *PLoS ONE, 12*(1), e0170754. https://doi.org/10.1371/journal.pone.0170754

Bassler, H. (2024). *In some parts of Canada, the 2023 fires never ended*. Wildfire Today. https://wildfiretoday.com/2024/01/26/in-some-parts-of-canada-the-2023-fires-never-ended/

Bauwens, J., & Naturale, A. (2017). The role of social work in the aftermath of disasters and traumatic events. *Clinical Social Work Journal, 45*(2), 99–101. https://doi.org/10.1007/s10615-017-0623-8

Boesecker, M., McDermott, J., & Condon, B. (2023). *An old car tire, burnt trees and a utility pole may be key in finding how the Maui wildfire spread*. The Associated Press. https://apnews.com/article/wildfires-maui-hawaii-electricity-utility-investigation-d9e1e84d5cc643d6bb2a2e362fe58ba0

Breslauer, G. W., & Breslauer, K. J. (2023). Political science meets physical science: The shared concept of stability. *PNAS Nexus. 2*(12). pgad401. https://doi.org/10.1093/pnasnexus/pgad401

Burns, M. (2022). *The 1/9 debris flow was not so rare*. Santa Barbara Independent. https://www.independent.com/2022/05/19/the-1-9-debris-flow-was-not-so-rare/

Carnicer, J., Alegria, A., Giannakopoulos, C., Di Giuseppe, F., Karali, A., Koutsias, N., Lionello, P., Parrington, M., & Vitolo, C. (2022). Global warming is shifting the relationships between fire weather and realized fire-induced CO_2 emissions in Europe. *Scientific Reports, 12*, 10365. https://doi.org/10.1038/s41598-022-14480-8

Castañer, X., & Oliveira, N. (2020). Collaboration, coordination, and cooperation among organizations: Establishing the distinctive meanings of these terms through a systematic literature review. *Journal of Management, 46*(6), 965–1001. https://doi.org/10.1177/0149206320901565

Cooper, J. A., & Pilkey, O. H. (2004). Sea-level rise and shoreline retreat: Time to abandon the Bruun rule. *Global and Planetary Change., 43*(3–4), 157–171. https://doi.org/10.1016/j.gloplacha.2004.07.001

Coronese, M., Lamperti, F., Keller, K., Chiaromonte, F., & Roventini, A. (2019). Evidence for sharp increase in the economic damages of extreme natural disasters. *Proceedings of the National Academy of Sciences of the United States of America, 116*(43), 21450–21455. https://doi.org/10.1073/pnas.1907826116

Dangendorf, S., Hendricks, N., Sun, Q., Klinck J., Ezer, T., Frederikse, T., Calafat, F. M., Wahl, T. & Törnqvist, T. E. (2023). Acceleration of U.S. Southeast and Gulf coast sea-level rise amplified by internal climate variability. *Nature Communications, 14*, 1935. https://doi.org/10.1038/s41467-023-37649-9

De Deckker, P. (2016). The Indo-Pacific warm pool: Critical to world oceanography and world climate. *Geosciences Letters, 3*, 20. https://doi.org/10.1186/s40562-016-0054-3

Diaz, R. J., & Rosenberg, R. (2011). Introduction to environmental and economic consequences of hypoxia. *International Journal of Water Resources Development, 27*(1), 71–82. https://doi.org/10.1080/07900627.2010.531379

Ekstrom, J. A., Moore, S. K., & Klinger, T. (2020). Examining harmful algal blooms through a disaster risk management lens: A case study of the 2015 U.S. West Coast domoic acid event. *Harmful Algae, 94*, 101740. https://doi.org/10.1016/j.hal.2020.101740

EPA. (2024). *Climate change impacts.* US Environmental Protection Agency. https://www.epa.gov/climateimpacts

Etheridge, S. M. (2010). Paralytic shellfish poisoning: Seafood safety and human health perspective. *Toxicon, 56*(2), 108–122. https://doi.org/10.1016/j.toxicon.2009.12.013

European Commission. (2022). *Drought in Europe: August 2022, GDO analytical report.* Joint Research Centre, Publications Office of the European Union. https://data.europa.eu/doi/10.2760/264241

Evans, M. & AccuWeather. (2012). Earth's strongest, most massive storm ever. *Scientific American.* https://www.scientificamerican.com/article/earths-strongest-most-massive-storm-ever/

Ezer, T. (2019). Regional differences in sea level rise between the Mid-Atlantic Bight and the South Atlantic Bight: Is the Gulf stream to blame? *Earth's Future, 7*(7), 771–783. https://doi.org/10.1029/2019EF001174

Ezer, T. (2020). Analysis of the changing patterns of seasonal flooding along the U.S. East Coast. *Ocean Dynamics, 70*(2), 241–255. https://doi.org/10.1007/s10236-019-01326-7

Fanti, V., Ferreira, Ó., Kümmerer, V., & Loureiro, C. (2023). Improved estimates of extreme wave conditions in coastal areas from calibrated global reanalyses. *Communications Earth & Environment, 4*, 151. https://doi.org/10.1038/s43247-023-00819-0

Fasullo, J. T., Rosenbloom, N., Buchholz, R. R., Danabasoglu, G., Lawrence, D. M., & Lamarque, J.-F. (2021). Coupled climate responses to recent Australian wildfire and COVID-19 emissions anomalies estimated in CESM2. *Geophysical Research Letters, 48*, e2021GL093841. https://doi.org/10.1029/2021GL093841

Federal Highway Administration (2016). *Living shoreline along coastal roadways exposed to sea level rise: Shore road in Brookhaven, New York,* FHWA-HEP-17–016. U.S. Department of Transportation.

Free, C. M., Anderson, S. C., Hellmers, E. A., Muhling, B. A., Navarro, M. O., Richerson, K., Rogers, L. A., Satterthwaite, W. H., Thompson, A. R., Burt, J. M., Gaines, S. D., Marshall, K. N., White, J. W., & Bellquist, L. F. (2023). Impact of the 2014–2016 marine heatwave on US and Canada West Coast fisheries: Surprises and lessons from key case studies. *Fish and Fisheries, 24*, 652–674. https://doi.org/10.1111/faf.12753

Garcia, K., & Castleman, T. (2023). *What are sneaker waves? Heart-stopping video shows how powerful surges can be dangerous.* Los Angeles Times. https://www.latimes.com/california/story/2023-12-29/what-are-sneaker-waves-heart-stopping-video-shows-how-powerful-surges-can-be-dangerous

Gemmrich, J., & Cicon, L. (2022). Generation mechanism and prediction of an observed extreme rogue wave. *Scientific Reports, 12*, 1718. https://doi.org/10.1038/s41598-022-05671-4

Goss, M., Swain, D. L., Abatzoglou, J. T., Sarhadi, A., Kolden, C. A., Williams, A. P., & Diffenbaugh, N. S. (2020). Climate change is increasing the likelihood of extreme autumn wildfire conditions across California. *Environmental Research Letters, 15*(9), 094016. https://doi.org/10.1088/1748-9326/ab83a7

Hannoun, D., & Tietjen, T. (2022). Lake management under severe drought: Lake Mead, Nevada/Arizona. *Journal of the American Water Resource Association, 59*(2), 416–428. https://doi.org/10.1111/1752-1688.13090

Hauer, M. E., Hardy, D., Kulp, S. A., Mueller, V., Wrathall, D. J., & Clark, P. U. (2021). Assessing population exposure to coastal flooding due to sea level rise. *Nature Communications, 12*, 6900. https://doi.org/10.1038/s41467-021-27260-1

Hart, G. F., & Coleman, J., (2005). *The world deltas database framework*. Louisiana State University. https://www.geol.lsu.edu/WDD/

Herren, L.W., Brewton, R. A., Wilking, L. E., Tarnowski, M. E., Vogel, M. A., & Lapointe, B. E., (2021). Septic systems drive nutrient enrichment of groundwaters and eutrophication in the urbanized Indian River Lagoon, Florida, *Marine Pollution Bulletin, 172*, 112928. https://doi.org/10.1016/j.marpolbul.2021.112928

Herring, S. C., Hoerling, M. P., Peterson, T. C., & Stott, P. A. (Eds.). (2014). Explaining extreme events of 2013 from a climate perspective. *Bulletin of the American Meteorological Society, 95*(9), S1–S96. https://doi.org/10.1175/1520-0477-95.9.S1.1

Hückstädt, M. (2023). Ten reasons why research collaborations succeed—a random forest approach. *Scientometrics, 128*, 1923–1950. https://doi.org/10.1007/s11192-022-04629-7

IFRC. (2020). *World Disasters Report 2020: Come heat or high water*. International Federation of Red Cross and Red Crescent Societies. https://www.ifrc.org/sites/default/files/2021-05/20201116_WorldDisasters_Full.pdf

Inman, D. L., & Dolan, R. (1989). The outer banks of North Carolina: budget of sediment and inlet dynamics along a migrating barrier island system. *Journal of Coastal Research, 5*(2), 193–237. https://www.jstor.org/stable/4297525

Institute for Economics & Peace. (2020). *Ecological threat register 2020: Understanding ecological threats, resilience and peace*. Institute for Economic & Peace. http://visionofhumanity.org/reports

Internal Displacement Monitoring Centre (n.d.). *Displacement, disasters and climate change*. Internal Displacement Monitoring Centre. https://www.internal-displacement.org/focus-areas/Displacement-disasters-and-climate-change/

IPCC. (2013). *Climate change 2013: The physical science basis. Contribution of working group I to the fifth assessment report of the intergovernmental panel on climate change* [Stocker, T.F., D. Qin, G.-K. Plattner, M. Tignor, S.K. Allen, J. Boschung, A. Nauels, Y. Xia, V. Bex and P.M. Midgley (eds.)]. Cambridge University Press. https://www.ipcc.ch/site/assets/uploads/2017/09/WG1AR5_Frontmatter_FINAL.pdf

Jaschik, K. (2014). Small states and international politics: Climate change, the Maldives and Tuvalu. *International Politics., 51*, 272–293. https://doi.org/10.1057/ip.2014.5

Juliano, T. W., Szasdi-Bardales, F., Lareau, N. P., Shamsaei, K., Kosović, B., Elhami-Khorasani, N., James, E. P., & Ebrahimian, H. (2024). Brief communication: The Lahaina fire disaster—how models can be used to understand and predict wildfires. *Natural Hazards and Earth System Sciences, 24*(1), 47–52. https://doi.org/10.5194/nhess-24-47-2024

Khojasteh, D., Haghani, M., Nicholls, R. J., Moftakhari, H., Sadat-Noori, M., Mach, K. J., Fagherazzi, S., Vafeidis, A. T., Barbier, E., Shamsipour, A., & Glamore, W. (2023). The evolving landscape of sea-level rise science from 1990 to 2021. *Communications Earth & Environment, 4*, 257. https://doi.org/10.1038/s43247-023-00920-4

Klotzbach, P. J., Wood, K. M., Bell, M. M., Blake, E. S., Bowen, S. G., Caron, L.-P., Colling, J. M., Gibnery, E. J., Schreck, C. J., III., & Truchelut, R. E. (2022). A hyperactive end to the Atlantic hurricane season October–November 2020. *Bulletin of the American Meteorological Society, 103*(1), E110–E128. https://doi.org/10.1175/BAMS-D-20-0312.1

Lacovelli, D., & Vasquez, T. (1998). Supertyphoon Tip. *Mariners Weather Log, 42*(2), 4–8. https://www.vos.noaa.gov/MWL/aug1998.pdf

Lander, B. (2022). *Ten years after Sandy: Barriers to resilience*. New York City Comptroller. https://comptroller.nyc.gov/wp-content/uploads/documents/Ten-Years-After-Sandy.pdf

Lanier, A. L., Drabik, J. R., Heikkila, T., Bolson, J., Sukop, M. C., Watkins, D. W., Rehage, J., Mirchi, A., Engel, V., & Letson, D. (2018). Facilitating integration in interdisciplinary research: Lessons from a South Florida water, sustainability, and climate project. *Environmental Management, 62*, 1025–1037. https://doi.org/10.1007/s00267-018-1099-1

Lee, J., Perera, D., Glickman, T., & Taing, L. (2020). Water-related disasters and their health impacts: A global review. *Progress in Disaster Science., 8*, 100123. https://doi.org/10.1016/j.pdisas.2020.100123

Lehmkuhl, F., Schüttrumpf, H., Schwarzbauer, J., Brüll, C., Dietze, M. Letmathe, P., Völker, C., & Hollert, H. (2022). Assessment of the 2021 summer flood in Central Europe. *Environmental Sciences Europe, 34* (107). https://doi.org/10.1186/s12302-022-00685-1

Lin, J-Y, Chao, J-C., & Lin, C-S. (2022). Evaluation of community resilience to debris-flow disasters: A case study of Nantou, Taiwan. *International Journal of Disaster Risk Reduction, 81*(2), 103258. https://doi.org/10.1016/j.ijdrr.2022.103258

Lincke, D., & Hinkel, J. (2021). Coastal migration due to 21st century sea-level rise, *Earth's Future. 9*(5), e2020EF001965. https://doi.org/10.1029/2020EF001965

Lindsey, R. (2022). *Climate change: Global Sea Level.* NOAA Climate.gov. https://www.climate.gov/news-features/understanding-climate/climate-change-global-sea-level

Liu, Y., Kerkering, H., & Weisberg, R. H. (Eds.) (2015). *Coastal ocean observing systems.* Elsevier. https://doi.org/10.1016/C2014-0-01713-3

Longshore, D. (2007). *Encyclopedia of hurricanes, typhoons, and Cyclones* (2nd ed.). Facts on File Inc.

Lund, J. R. (2015). Integrating social and physical sciences in water management. *Water Resources Research, 51*(8), 5905–5918. https://doi.org/10.1002/2015WR017125

Manoj, J. M., Sautter, B., Ariffin, E. H., Menier, D., Mu, R., Siddiqui, N. A., Delanoe, H., Del Estal, N., Traoré, K., & Gensac, E. (2020). Total vulnerability of the littoral zone to climate change-driven natural hazards in north Brittany, France. *Science of the Total Environment, 706*, 135963. https://doi.org/10.1016/j.scitotenv.2019.135963

Marra, J. (2015). *Understanding climate: Billy Sweet and John Marra explain nuisance floods.* NOAA. https://www.climate.gov/news-features/understanding-climate/understanding-climate-billy-sweet-and-john-marra-explain

Melet, A., van de Wal, R., Amores, A., Arns, A., Chaigneau, A. A., Dinu, I., Haigh, I. D., Hermans, T. H. J., Lionello, P., Marcos, M., Meier, H. E. M., Meyssignac, B., Palmer, M. D., Reese, R., Simpson, M. J. R., & Slangen, A. (2023). Sea level rise in Europe: Observations and projections. *State of the Planet Discussions.* [preprint] https://doi.org/10.5194/sp-2023-36

Moore, A. (2022). *Climate change is making wildfires worse—Here's how.* North Carolina State University. https://cnr.ncsu.edu/news/2022/08/climate-change-wildfires-explained/

Morea, H., & Samanta, S. (2020). Multi-criteria decision approach to identify flood vulnerability zones using geospatial technology in the Kemp-Welch catchment, Central Province, Papua New Guinea. *Applied Geomatics, 12*, 427–440. https://doi.org/10.1007/s12518-020-00315-6

Mueting, A., Bookhagen, B., & Strecker, M. R. (2021). Identification of debris-flow channels using high-resolution topographic data: A case study in the Quebrada del Toro, NW Argentina. *Journal of Geophysical Research: Earth Surface, 126*, e2021JF006330. https://doi.org/10.1029/2021JF006330

Mulligan, T. D., Taylor, K., & DeLeo, R. A. (2019). Politics and policies for managing natural hazards. *Natural Hazard Science.* https://doi.org/10.1093/acrefore/9780199389407.013.314

Munkes, B., Löptien, U., & Dietze, H. (2021). Cyanobacteria blooms in the Baltic sea: A review of models and facts. *Biogeosciences, 18*, 2347–2378. https://doi.org/10.5194/bg-18-2347-2021

Nairn, J., Beaty, M., & Varghese, B. M. (2021). Australia's black summer heatwave impacts. *Australian Journal of Emergency Management, 36*(1), 17–20. https://knowledge.aidr.org.au/media/8401/ajem_06_2021-01.pdf

Nanzetti, S. (2018). Mathematical modeling of rogue waves: A survey of recent and emerging mathematical methods and solutions. *Axioms, 7*(2), 42. https://doi.org/10.3390/axioms7020042

Naqvi, S., & Wajih, A. (2021). Deoxygenation in marginal seas of the Indian Ocean. *Frontiers in Marine Science, 8*, Article 624322. https://doi.org/10.3389/fmars.2021.624322

Newman, R., & Noy, I. (2023). The global costs of extreme weather that are attributable to climate change. *Nature Communications., 14*, 6103. https://doi.org/10.1038/s41467-023-41888-1

Nichols, C. R., & Raghukumar, K. (2020). *Marine environmental characterization.* Springer Cham. https://doi.org/10.1007/978-3-031-02490-0

NWCG. (2020). *Glossary of wildland fire terminology.* National Wildfire Coordinating Group. https://www.nwcg.gov/publications/pms205/nwcg-glossary-of-wildland-fire-pms-205

Palmer, T., & Stevens, B. (2019). The scientific challenge of understanding and estimating climate change. *Proceedings of the National Academy of Sciences of the United States of America., 116*(49), 24390–24395. https://doi.org/10.1073/pnas.1906691116

Patel, K. (2022). *Australia hits 123 degrees, tying highest temperature on record in Southern Hemisphere.* The Washington Post. https://www.washingtonpost.com/weather/2022/01/13/australia-southern-hemisphere-temperature-record/

Paterson, P. (2019). Climate change is coming for Annapolis. *USNI Proceedings. 145*(10). https://www.usni.org/magazines/proceedings/2019/october/climate-change-coming-annapolis

Pennington, C. (n.d.). *Holbeck Hall landslide, Scarborogh.* British Geological Survey. https://www.bgs.ac.uk/discovering-geology/maps-and-resources/office-geology/holbeck-hall/

Petryk, V. (2023). *Natural disasters 2022: Tragic losses & lessons learnt.* EOS Data Analytics. https://eos.com/blog/natural-disasters-2022/

Pietrafesa, L. J., Zhang, H., Bao, S., Gayes, P. T., & Hallstrom, J. (2019). Coastal flooding and inundation and inland flooding due to downstream blocking. *Journal of Marine Science and Engineering, 7*(10), 336. https://doi.org/10.3390/jmse7100336

Plough, A. L., & Chandra, A. (2015). *What Katrina taught us about community resilience.* RAND. https://www.rand.org/pubs/commentary/2015/09/what-hurricane-katrina-taught-us-about-community-resilience.html

Qin, B., Deng, J., Shi, K., Wang, J., Brookes, J., Zhou, J., Zhang, Y., Zhu, G., Paerl, H. W., & Wu, L. (2021). Extreme climate anomalies enhancing cyanobacterial blooms in eutrophic Lake Taihu, China, *Water Resources Research, 57*, e2020WR029371. https://doi.org/10.1029/2020WR029371

Rehak, J. A. (2024). *We live in the water: Climate, aging, and socioecology on Smith Island.* Johns Hopkins University Press.

Rodell, M., & Li, B. (2023). Changing intensity of hydroclimatic extreme events revealed by GRACE and GRACE-FO. *Nature Water, 1*, 241–248. https://doi.org/10.1038/s44221-023-00040-5

Rosen, B. H., Davis, T. W., Gobler, C. J., Kramer, B. J., & Loftin, K. A. (2017). *Cyanobacteria of the 2016 Lake Okeechobee waterway harmful algal bloom: U.S.* Geological Survey Open-File Report 2017–1054. https://doi.org/10.3133/ofr20171054

SECOORA. (2022). *Eyes on Ian: Data resources. Southeast Coastal ocean observing regional association.* https://secoora.org/eyes-on-ian-data-resources/

Shaw, J. E., & Zamorano, M. (2020). Saltwater interface monitoring and mapping program, Technical Publication WS-58, South Florida Water Management District. https://www.sfwmd.gov/sites/default/files/documents/ws-58_swi_mapping_report_final.pdf

Sidik, S. M. (2020). Feedback loops of fire activity and climate change in Canada. *Eos, 101*. https://doi.org/10.1029/2020EO152247

Sova, C. & Zembilci, E. (2023). *Dangerously hungry: The link between food insecurity and conflict.* Center for Strategic & International Studies. https://www.csis.org/analysis/dangerously-hungry-link-between-food-insecurity-and-conflict

Stock, J. D., & Dietrich, W. E. (2006). Erosion of steepland valleys by debris flows. *GSA Bulletin, 118*(9–10), 1125–1148. https://doi.org/10.1130/B25902.1

Stevens-Rumann, C. S., Kemp, K. B., Higuera, P. E., Harvey, B. J., Rother, M. T., Donato, D. C., Morgan, P., & Veblen, T. T. (2018). Evidence for declining forest resilience to wildfires under climate change. *Ecology Letters, 21*(2), 243–252. https://doi.org/10.1111/ele.12889

Strauss, B. H., Orton, P. M., Bittermann, K., Buchanan, M. K., Gilford, D. M., Kopp, R. E., Kulp, S., Massey, C., de Moel, H., & Vinogradov, S. (2021). Economic damages from Hurricane Sandy attributable to sea level rise caused by anthropogenic climate change. *Nature Communications, 12*, 2720. https://doi.org/10.1038/s41467-021-22838-1

Susskind, J., Schmidt, G. A., Lee, J. N., & Iredell, L. (2019). Recent global warming as confirmed by AIRS. *Environmental Research Letters, 14*(4), 044030. https://doi.org/10.1088/1748-9326/aafd4e

Tonn, G., & Czajkowski, J. (2022). Evaluating the risk and complexity of pluvial flood damage in the U.S. *Water Economics and Policy, 8*(3), 2240002. https://doi.org/10.1142/S2382624X22400021

Tosi, L., Da Lio, C., Bergamasco, A., Cosma, M., Cavallina, C., Fasson, A., Viezzoli, A., Zaggia, L., & Donnici, S. (2022). Sensitivity, hazard, and vulnerability of farmlands to saltwater intrusion in low-lying coastal areas of Venice, Italy. *Water, 14*(1), 64. https://doi.org/10.3390/w14010064

Törnqvist, T. E., Jankowski, K. L., Li, Y-X., & González, J. L. (2020). Tipping points of Mississippi Delta marshes due to accelerated sea-level rise. *Science Advances, 6*(21), eaaz5512. https://doi.org/10.1126/sciadv.aaz5512

Trotman A., Joyette, A., van Meerbeeck, C., Mahon, R., Cox, S., Cave, N., & Farrell, D. (2018). Drought risk management in the Caribbean community: Early warning information and other risk reduction considerations. In D. Wilhite & R. Pulwarty, (Eds.). *Drought and water crises.* CRC Press. https://doi.org/10.1201/b22009-24

Toreti, A., Bavera, D., Acosta Navarro, J., Cammalleri, C., de Jager, A., Di Ciollo, C., Hrast Essenfelder, A., Maetens, W., Magni, D., Masante, D., Mazzeschi, M., Niemeyer, S., & Spinoni, J. (2022). *Drought in Europe August 2022*, JRC130493. Office of the European Union, https://doi.org/10.2760/264241

United Nations Office for Disaster Risk Reduction. (2021). *GAR Special Report on Drought 2021.* Geneva. https://www.droughtmanagement.info/literature/UNGAR_Specia_Report_on_Drought_2021.pdf

Van Wyk, H., Eisenberg, J. N. S., & Brouwer, A. F. (2023). Long-term projections of the impacts of warming temperatures on Zika and dengue risk in four Brazilian cities using a temperature-dependent basic reproduction number. *PLoS Neglected Tropical Diseases, 17*(4), e0010839. https://journals.plos.org/plosntds/article?id=10.1371/journal.pntd.0010839

Veeramony, J., Li, C., Sheremet, A., Myers, E. P., III., Xia, M., & Mitchell, S. (2020). Estuarine and coastal natural hazards: An introduction and synthesis. *Estuarine Coastal and Shelf Science., 237*, 106654. https://doi.org/10.1016/j.ecss.2020.106654

Vinogradova, N., & Hamlington, B. (2022). Sea level science and applications support coastal resilience. *Eos, 103.* https://doi.org/10.1029/2022EO220301

Warrick, R. A., Azizul Hoq Bhuiya, A. K., Mitchell, W. M., Murty, T. S., & Rasheed, K. B. S. (1996). Sea-level Changes in the Bay of Bengal. In R. A. Warrick, R.A. & Q. K. Ahmad (Eds.), *The implications of climate and sea–level change for Bangladesh* (pp. 97–142). Springer. https://doi.org/10.1007/978-94-009-0241-1_3

Westerling, A. L., Hidalgo, H., Cayan, D. R., & Swetnam, T. W. (2006). Warming and earlier spring increase Western U.S. forest wildfire activity. *Science, 313*(5789), 940–943. https://www.science.org/doi/10.1126/science.1128834

Withers, P. J. A., Jordan, P., May, L., Jarvie, H. P., & Deal, N. E. (2014). Do septic tank systems pose a hidden threat to water quality? *Frontiers in Ecology and the Environment, 12*(2), 123–130. https://doi.org/10.1890/130131

Wright, L. D., & Nichols, C. R. (Eds.) (2019). *Tomorrow's coasts: complex and impermanent, coastal research library*, 27, Springer.

World Bank Group. (2021). *Papua New Guinea. Climate change knowledge portal.* https://bit.ly/3RsSZar

WMO. (2019). *Vision for the WMO integrated global observing system in 2040, WMO-No. 1243.* World Meteorological Organization. https://library.wmo.int/doc_num.php?explnum_id=10278

WMO. (2020). *World's deadliest tropical cyclone was 50 years ago.* World Meteorological Organization. https://web.archive.org/web/20231218172627/https://public-old.wmo.int/en/media/news/world%E2%80%99s-deadliest-tropical-cyclone-was-50-years-ago

WMO. (2023). *Storm Daniel leads to extreme rain and floods in Mediterranean, heavy loss of life in Libya.* World Meteorological Organization. https://wmo.int/media/news/storm-daniel-leads-extreme-rain-and-floods-mediterranean-heavy-loss-of-life-libya

Yan, J., Yang, P., Wang, B.,Wu, S., Zhao, M., Zheng, X., Wang, Z., Zhang, Y., & Fan, C. (2024). Green roof systems for rainwater and sewage treatment. *Water, 16,* 2090. https://doi.org/10.3390/w16152090

Zarillo, G. A. (2023). Inter-annual sea level change and transgression along a barrier Island coast. *Frontiers in Environmental Science, 11,* 1107458. https://doi.org/10.3389/fenvs.2023.1107458

Zhu, L., & Quiring, S. M. (2022). Exposure to precipitation from tropical cyclones has increased over the continental United States from 1948 to 2019. *Communications Earth & Environment, 3,* 312. https://doi.org/10.1038/s43247-022-00639-8

Planning for and Managing Evolving Future Risks

3

The coastal ocean system is an important mechanism, which connects the continents to the ocean while moderating air and water temperatures. Radiation for the Sun is absorbed by the ocean and then the ocean distributes excess heat from the equator towards the poles. This transfer is done by permanent ocean currents such as the Gulf Stream, which acts like a conveyor belt, transporting warm water and precipitation from the equator toward the poles and deep ocean currents carry dense cold water from the poles back to the tropics. Thus, ocean currents regulate global climate, helping to counteract the uneven distribution of solar radiation reaching Earth's surface. Without currents in the ocean, regional temperatures would be more extreme—super hot at the equator and frigid toward the poles. For these reasons, observing and predicting sea surface temperatures is very important. Changes in sea surface temperatures are observed and mapped by agencies such as the European Space Agency, NASA, and NOAA.

3.1 Rising Sea Temperatures and Evolving Storm Threats

Rising sea surface temperatures (SST) are increasing coastal hazards faster than previously expected (Wright & Thom, 2023). Impacts include rapid tropical storm development, more "compound events" involving storm surges combined with torrential rainfall, bleaching and deaths of coral reefs, hypoxia of coastal and estuarine waters, the accelerated melting of sea ice; weakening of the AMOC; increases in harmful algal blooms and pathogens in estuaries; and lingering flood waters. Record Northern Hemisphere SST in the summer of 2023 (Fig. 3.1) significantly exceeded predictions and are expected to be even higher in 2024. Regional, global, and local planning for future coastal resilience

© The Author(s), under exclusive license to Springer Nature Switzerland AG 2025
C. Reid Nichols et al., *Integrated Coastal Resilience*, Synthesis Lectures on Ocean Systems Engineering, https://doi.org/10.1007/978-3-031-68153-0_3

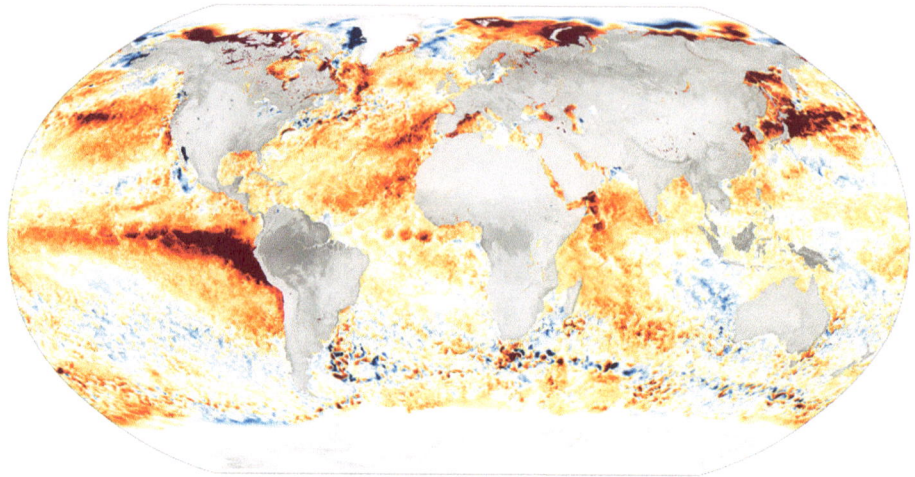

Fig. 3.1 Global sea surface temperature anomalies on 21 August 2023. Dark red indicates temperatures more than 3 °C above average. Orange areas are 1–2 °C above average *Source* NASA Earth Observatory, based on data from the Multiscale Ultrahigh Resolution Sea Surface Temperature (MUR SST) project. Record high temperatures prevailed for four consecutive months

necessitates near term improvements of numerical models of SST along with high resolution observations of the ocean and atmospheric factors responsible for warming seas and bays. Evolution of improved and diverse locally specific strategies for mitigating the impacts of hot sea surfaces on coastal erosion, inundation, ecosystem degradation, and coral bleaching should become a priority for coastal scientists and managers. The harmful impacts of rising ocean temperatures include:

- rising global sea level;
- increasing size and intensity of the tropical storms that bring destructive waves and storm surges;
- accelerating melting of sea ice;
- bleaching and death of protective coral reefs;
- weakening of the AMOC and the Gulf Stream;
- reductions in dissolved oxygen;
- increases in harmful algal blooms; and
- transport of pathogens in lingering flood waters and estuaries.

Scientists and engineers use a variety of in situ ocean instruments to measure ocean heat content at different depths. Examples include Argo floats (see https://argo.ucsd.edu/) and ocean gliders (see https://www.oceangliders.org/).

Although rising sea levels are expected to continue to cause inundation and land loss of low elevation coastal zones into the distant future, it is not sea level rise but compound floods from combined storm surge and torrential rains that coastal dwellers should fear most. Tropical cyclones are now developing faster and becoming larger and more intense because of elevated sea surface temperatures. The joint probability of severe storm surge and pluvial flooding coinciding in U.S. coastal cities has increased significantly over the past century (Wahl et al., 2015). In cases where rivers are nearby, fluvial flooding further exacerbates the severity of inundation. The International Human Dimensions Program on Global Environmental Change points out that by midcentury, the flood risk to large coastal cities will have increased by nine-fold relative to the present day. According to NOAA's Office of Coastal Management, inundation events are the dominant causes of natural-hazard-related deaths in the U.S. and are also the most frequent and costly of the natural hazards affecting the nation. The spatially extensive analyses of meteorological and coastal oceanographic data show that the occurrence of compound coastal flooding has been increasing over the past several decades (EPA, 2023). Unanticipated flooding in recent years was caused by intense rains that accompanied Hurricanes Harvey, Irma, Maria, Florence, Dorian, Ida and, most recently, Ian in September 2022. Hurricane Irma (2017) was the longest-lived Category 5 hurricane on record in the Atlantic Basin since the satellite era began in 1966, maintaining that intensity for 3.25 days, according to Klotzbach et al. (2022). Flooding that followed was amplified by a combination of drivers such as storm surge, intense rainfall, and high river discharge. The ensuing compound floods lead to greater impact than when the flooding occurs as the result of a single driver.

There are some serious lessons from Ian that should be heeded by planners and coastal residents in the future. What follows is a summary of some of Ian's behaviors that departed from "traditional" behaviors (see Fig. 3.2). When Hurricane Ian made landfall on Florida's Gulf coast near Ft. Myers on September 28, 2022, it was the most recent, and one of the five most severe, storms to hit this coast. Ian began as a tropical wave that moved off the coast of West Africa and eventually entered the Caribbean on September 21, 2022. It became a tropical depression on September 23 and a tropical storm the next day. Then within only 24 h it strengthened from a tropical storm to a major Category 3 hurricane before making landfall on western Cuba and then tracking northward to Florida over the warm (31.11° C) Gulf of Mexico and ultimately making its Florida landfall as a huge Category 4 storm, nearly a Category 5. Like several other storms that have impacted the U.S. coasts in recent years, the rapid intensification and large size of Ian was almost certainly caused by the higher-than-normal sea surface temperatures. These effects are expected to become more severe over the coming years.

Fig. 3.2 Hurricane Ian at time of landfall near Ft. Myers FL. Wednesday, Sept. 28, 2022. Florida's death toll was 150 people (Bucci et al., 2023)

On September 26, two days prior to Ian's ultimate landfall, the predicted "cone of uncertainty" for the storm's predicted future path, showed the most likely region of landfall to be in the "Big Bend" near the northern extremity of the Nature Coast[1] and well to the north of Tampa Bay. The storm track is depicted in Fig. 3.3. By the following day, Tuesday, September 27, the cone of the possible track of Ian extended from the Big Bend in the north to Charlotte Harbor in the south with the centerline passing over Tampa Bay where the storm surge was then forecast to potentially exceed 10 ft (see Fig. 3.4). However, as the storm crossed the western end of Cuba, the track began a turn to the right and in a more easterly direction. Ian ultimately made landfall well to the south of

---

[1] Nature Coast refers to eight counties (Wakulla, Jefferson, Taylor, Dixie, Levy, Citrus, Hernando, and Pasco) that abut the Gulf of Mexico in Florida.

the Tampa Bay area with the eye passing over Ft. Myers. The counterclockwise circulation caused onshore winds to the south of the eye and offshore winds to the north. The National Hurricane Center (NHC) predicted storm surge about two hours before landfall were for surge heights south of the eye to exceed 2.74 m above ground level as far south as the Everglades. The storm surge caused widespread devastation to the south of the eye but spared the Tampa Bay region and the Nature Coast to the north of Tampa because the offshore winds suppressed water level by pushing water out of the bay and out to sea along the coast to the north.

The rapid and unexpected flooding of the regions affected by the storm surge contributed to the loss of approximately 148 lives, in part because the warnings came too late to allow effective evacuation. The storm track turned out to be within the southern extremity of the NHC forecast cone of uncertainty but well removed from the "centerline" path where landfall was expected and prepared for (see Figs. 3.2 and 3.3). In the future, warnings and calls for evacuation should probably embrace the entire cone of uncertainty and not focus on the center line. A lesson learned from this storm is that storms are now more

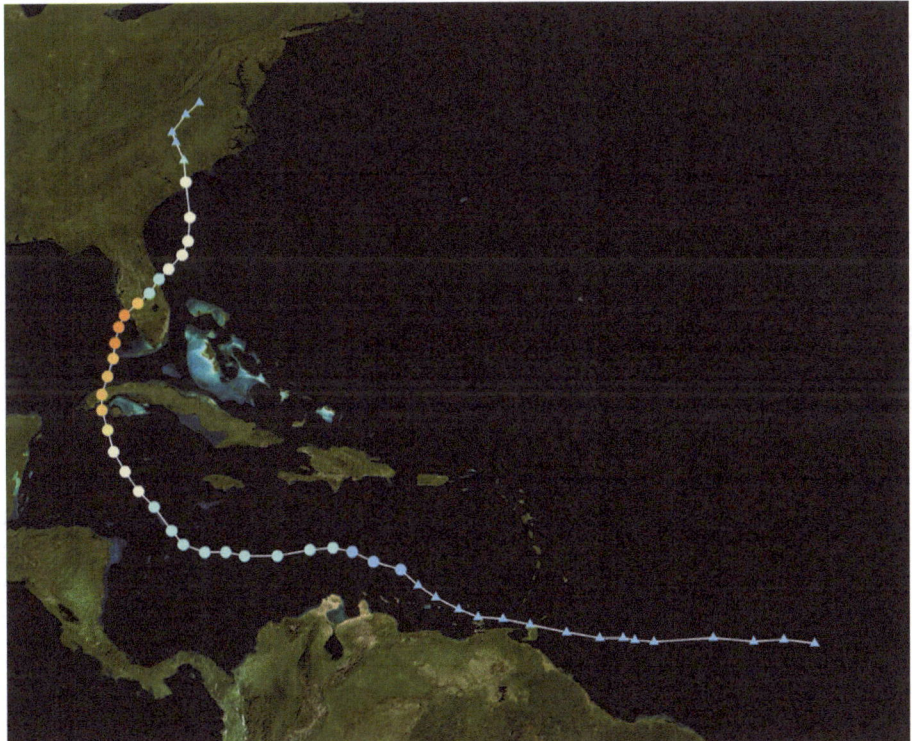

Fig. 3.3 Hurricane Ian storm track from 23–30 September 2022. Ian made landfall in a region extremely vulnerable to storm surge (Bucci et al., 2023)

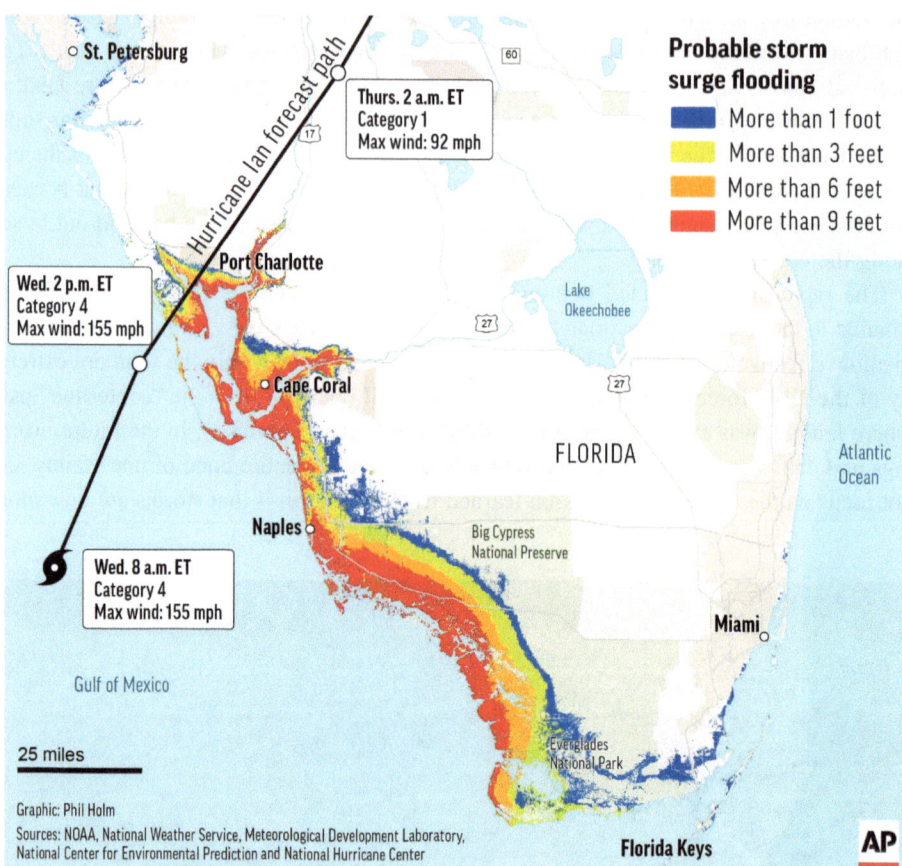

Fig. 3.4 Storm surge warning for the South Florida Coast on 28 September 2022 (*Source* NOAA National Weather Service). Ian's surge of at least 4.2 m (13.8 ft) along Fort Myers Beach is the highest surge ever recorded in Southwest Florida over the past 150 years

dangerous than in the recent past and they can change course and intensify very quickly. In order to improve resilience against storm surges, researchers who are improving forecasts need to be engaged in tests that include maintaining accurate data, realistic scenario planning, and cross-functional collaboration with stakeholders. An example success involves annual exercises that utilize NOAA's Aviation Weather Testbed (see https://testbed.aviationweather.gov/) to assess forecasting innovations with stakeholders that include weather forecasters, air traffic controllers, and pilots.

The ability of a community to recover from a flood event depends more on local factors than on regional or global factors. For example, drainage infrastructure determines how long contaminated flooding waters may linger in neighborhoods and urban areas. Lingering floodwaters elevate risk due to physical hazards, increase property damage, and prolong contact with disease agents, displaced animals, and toxic chemicals common in stormwater. The adverse consequences of flood events, especially coastal flooding, to human health have been evaluated by the World Health Organization (2003) and the U.S. Global Change Research Program (USGCRP, 2016). Public health professionals, including epidemiologists, must be involved in planning and post-storm recovery. The interconnections of water infrastructure (e.g. drainage), public health and sea level rise require careful consideration and complex modeling (Allen et al., 2021). Post-event recovery plans need to include detailed monitoring of water depth and water borne pathogens. It is possible that some of the monitoring could utilize trained "citizen scientists".

3.2 Coastal Morphodynamic Responses to Climate Change

"The shape of the coast and the processes that mold it change together as a complex system. There is constant feedback among the multiple components of the system, and when climate changes, all facets of the system change. Abrupt shifts to different states can also take place when certain tipping points are crossed" (Wright & Thom, 2023). The quote above is intended to communicate the fact that complex suites of feedback between the energetic forces that erode, flood or change the shapes of coastal lands must be understood before the full impacts of climate change can be understood. The morphodynamic approach to explaining coastal behavior has been widely deployed over the past several decades (a recent example is the recent multi authored book edited by Jackson and Short, 2020). Recent examples of climate change impacts on beach and barrier island erosion, river deltas, estuaries and bays, continental shelves, Arctic coasts, wetlands, coral reefs, and built infrastructure are referenced in the review by Wright and Thom (2023). One recent example of an analytical model for predicting over wash and erosion of sandy barrier islands is described by Nienhuis et al. (2021).

As reported by Wright and Thom (2023), Ranasinghe et al. (2023) applied a multiscale Probabilistic Coastline Recession (PCR) model to a swell-dominated beach near Sydney, Australia, and a storm-dominated Netherlands case of long-term recession caused by century long periods of sea level rise, storm waves, and storm surge. It was concluded that sea level rise likely plays the dominant role in the long-term recession of both types of beach regimes. Long-term changes in wave climate were predicted to have only marginal impacts on recession. However, it must be noted that these results did not take account of the recent intensification of tropical cyclonic storms. These modeling approaches provide a valuable tool to learn how dynamic coastal systems operate by testing hypotheses and studying the role of the various factors.

3.3 Increasing Risks for Deltaic Coasts and Low Elevation Coastal Zones

Organizations such as the European Geophysics Union have reported on deltas and their growing populations owing to factors such as fertile floodplains, easy access to the ocean, and abundant land. World-wide, more than 500 million people live on or near river deltas (Giosan et al., 2014). Many of these deltas support megacities with populations of over 10 million people. According to the European Space Agency (https://www.esa.int/ESA_Mul timedia/Images/2020/10/Ganges_Delta), the Ganges–Brahmaputra Delta in Bangladesh and India supports over 100 million inhabitants. Most of these people are impoverished and highly vulnerable. In comparison, the Mississippi Delta has a little more than 2 million inhabitants. Globally, there are about 11,000 deltas and about 17 of these deltas are occupied by at least a million or more people. These include the Ganges–Brahmaputra, Chao Phraya, Irrawaddy, Indus, Nile, Godavari, Niger, Pearl, Krishna, Limpopo, Mahanadi, Mekong, Tigris, Changjiang (Yangtze), Huanghe (Yellow), Parana, and Mississippi (Overeem & Syvitski, 2009). Owing to factors such as fertile floodplains and easy access to the ocean, deltas will probably continue to maintain high population densities and climate change threats.

The additive effects of land subsidence, sea level rise, severance of sediment supply by the damming of rivers, and the increasing severity of storms are causing accelerating land losses in a majority of deltas (Wright et al., 2019a, 2019b). The future resilience of communities that currently occupy deltaic lands will depend on expensive engineering adaptations such as seawalls, dikes and levees and in many cases on abandoning existing towns and moving to higher ground. Long range planning must include extensive use of complex systems models and high-performance computing facilities to predict future threats and evaluate the human consequences of alternative strategies. To monitor the changes in "disappearing deltas", NASA's Jet Propulsion Laboratory recently launched its "Delta-X" initiative to acquire and analyze high resolution airborne data focused on delta vulnerability and resilience (Hall, 2021). NASA is currently utilizing the rapidly receding Mississippi Delta as a testbed for refining their methodology.

According to Allison et al. (2016), the rates of Mississippi Delta subsidence are up to 18 mm/yr. When this subsidence is added to the projected rates of global sea level rise of between 8 mm/yr and 16 mm/yr, the total relative rate of sea level rise in coastal Louisiana will conceivably be between 26 mm/yr and 34 mm/yr or roughly up to 1 foot per decade. As pointed out earlier, recent studies (Turner et al. 2018; Törnqvist et al., 2020) have concluded that a tipping point for submergence of deltaic and coastal wetland surfaces beyond which "dry" land is replaced by open water is around 5 mm/yr. This tipping point is now exceeded by 3- to fivefold in much of Coastal Louisiana and the current rate of land loss is estimated to be over 41 km^2yr^{-1} and increasing. Figure 3.5 highlights the extent of wetlands of Coastal Louisiana as of 2017. The Louisiana Coastal Protection and Reclamation Authority (CPRA, 2017) projects that by 2050, without reclamation, most of the wetlands will have been replaced by open water (Fig. 3.6). However, CPRA's restoration plans, which include diversion of sediment-supplying tributaries will probably provide solutions in specific localities. A recent study of land surface elevations through-out China utilizing spaceborne synthetic aperture radar (Ao et al., 2024) revealed that China's coastal lands have been subsiding by 3–10 mm/yr and that by the end of the cen-tury as much as 26% of these surfaces could be below sea level. The subsidence is most severe around coastal megacities like Shanghai and Guangzhou because of groundwater extraction and the weight of high-rise buildings.

Subsidence is also occurring in most other deltas and at varying rates. Davydzenka et al., (2024, p. 3, Fig. 2.1) produced a global map of land subsidence based on machine learning.

While subsidence occurs naturally as sediments compact over time, human activities such as groundwater withdrawal, oil and gas extraction, sand mining, and the construction of flood barriers around rivers, all contribute to subsidence. In parts of China's Pearl River Delta, the rates are up to 15 mm/yr and with a projected maximum sea level rise of an additional 1.3 m by 2100, much of the existing land surface could be replaced by open water (Wright et al., 2019a, 2019b). Serious subsidence of as much as 40 mm/yr is taking place in the Venice lagoon (Allison et al., 2016). Coastal subsidence will increase the

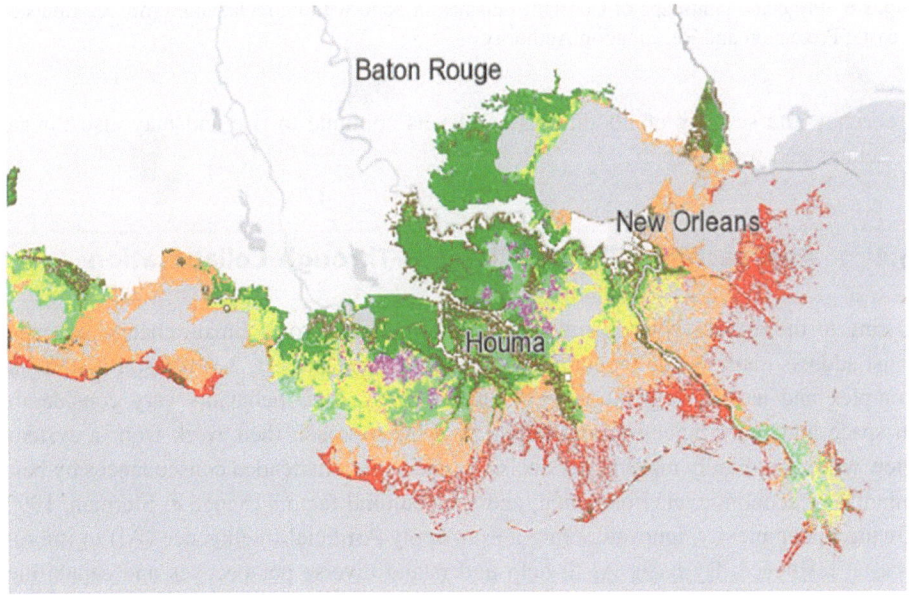

Fig. 3.5 Wetlands of Coastal Louisiana in 2017. (Dark green = forest and swamp; light green = freshwater marsh; yellow = fragile floating marsh; orange = brackish marsh; red = salt marsh; *Source* Louisiana Coastal Protection and Reclamation Authority)

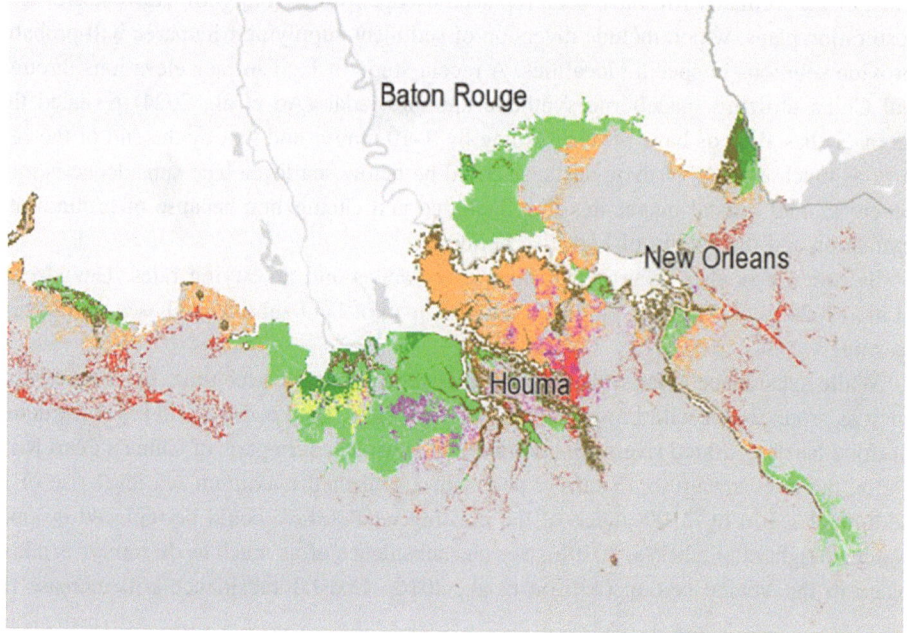

Fig. 3.6 Projected landscape of Coastal Louisiana in 2050 with no reclamation (*Source* Louisiana Coastal Protection and Reclamation Authority)

frequency and severity of flooding as sea levels continue to rise and may also threaten structures and water supply.

3.4 Approaching Climatic Changes Through Collaboration

Owing to the impact of environmental factors resulting from climate change, engineers must address many of the potential problems described in Sect. 2.1. Coastal systems are complex and have a large number of unique parts. Their behaviors vary considerably in space and time. Scientists and engineers who approach their work from a systemic view will undoubtedly make better decisions and avoid unattended consequences by being mindful of economic, environmental, and sociocultural factors (Senge & Sterman, 1992). Owing to complexity, innovators might even apply Artificial Intelligence (AI) to improve coastal resilience. By using AI to help understand diverse perspectives and capabilities, teams are provided with evidence that can improve decision making. Wilson and Daugherty (2018) described the potential of AI to enhance analytic and decision-making abilities and heighten creativity. Through the processing of large amounts of data and predictive analyses, AI can influence the coastal system's resilience.

Big data are critical to support more resource-efficient improvements in resilience, especially disaster recovery. Engineers must utilize the growing volume of historic data, real-time data from sensor networks, imagery, and model output on their projects (e.g., Benzhaf et al., 2022; Pollard et al., 2018; Sarker et al., 2020). Traditional approaches may apply multiple regression analysis techniques to consider environmental load evaluations based on work processes, input materials, and energy sources as described by Cho et al. (2017) during tunnel construction. Emerging capabilities such as AI can be used to extend water level time series to better understand extreme sea levels (Dubois et al., 2024). Information from stakeholders who will participate in the design of infrastructure or capabilities that help communities adapt to the effects of climate change is also important. Coastal data that support resilience initiatives are derived from satellites, aircraft, vessels, and monitoring systems to provide science and information to users from governments, universities, corporations, and communities, worldwide.

The USACE has sponsored collaborative initiatives to study coastal processes at locations such as the Field Research Facility (FRF; see the FRF data portal online at https://frfdataportal.erdc.dren.mil/). For example, the recent DUNEX (DUring Nearshore Event Experiment; Cialone et al., 2023; Straub et al., 2023) at the FRF applied multisource data to improve observational and predictive capabilities during storms such as Hurricane Larry (2021) while also supporting effective communication between scientists and the coastal community. Experiments that include citizen scientists suggest that big data analytics are fundamental to developing resilient coastal structures that are capable of withstanding extreme weather events and rising sea levels, as well as preventing and controlling releases of greenhouse gases. Coastal communities benefit when diverse public and private organizations collaborate to improve their coastal resilience.

Organizations such as the National Society for Professional Engineers recognize that attention to sustainable and resilient design practices is critical to the health of the planet and is an integral part of the practice of engineering. Many companies today apply these concepts by adopting standards of sustainability and social responsibility that relate to people, profits, and planet (the 3 P's), aka, the triple bottom line (Elkington, 2018). Michelin, for example, has achieved an equitable balance among human, economic and environmental issues. Their efforts are evidenced by (1) establishing projects with the World Wildlife Foundation that support an ecological reserve in Brazil, (2) ongoing investigations focused on sustainable rubber tree farming, (3) focusing research on the manufacturing of tires that have no impact on resources and biodiversity, and (4) operating net-zero CO_2 emission plants by 2050. Success in considering the 3 P's involves applications of stakeholder theory (Freeman & Dmytriyez, 2017) and building long-term partnerships. Engineering projects should build relationships and create value for all stakeholders, including government, business, and community partners.

In planning adaptation strategies for the next few decades, it is important for coastal engineers and managers to not rely on past experiences for their decisions. Climate change is already causing significant shifts in coastal behavior and the future may well hold some previously unforeseen challenges. The most significant of the anticipated future alterations of coastal risks are (1) changes in the rate of formation, intensity and size of tropical cyclones; (2) submergence and land loss in deltas and low elevation coastal zones; and (3) "tipping points" beyond which coastal systems can shift to completely new modes of behavior and equilibrium (Allen et al., 2021). Engineers contribute to resilience by using tools such as the USACE Sea Level Analysis Tool (SLAT; see https://climate. sec.usace.army.mil/slat/) to visualize observed sea level data and compare observations to projected sea level change while incorporating sea level rise rates into their coastal projects. Scientists, engineers, and policy makers must collectively analyze large datasets, integrate complex climate models, and develop simulations to stay in front of climate change impacts.

Many communities are also battling invasive species. Some communities have found that coastal ecosystems and cultural resources face threats from invasive species such as buffelgrass, which burns at high temperatures and helped to fuel wildfires on Maui during August 2023. Buffelgrass is native to Africa and is very flammable (Cornwall, 2022). While long periods of hot, dry weather elevate fire risks, an actual spark is required to start a grass or wildfire. Most are started by human activity or lightning. Wildfires might be prevented by the removal of dry grasses that are highly flammable around populated areas to include clearing defensible space around houses and ensuring viable evacuation plans. Communities can contribute to resilience by working with scientists and engineers. Government, academia, and industry are natural partners to share resources on projects aimed at improving coastal resilience at the community level. When a community has a robust network of experts, it is expanding its resources and resilience will improve. However, one of the most formidable challenges that scientists and engineers face in convincing communities to make adaptive changes and investments involves overcoming chronic misinformation and a distrust in scientists.

References

Allen, T., Behr, J., Bukvic, A., Calder, R. S. D., Caruson, K., Connor, C., D'Elia, C., Dismukes, D., Ersing, R., Franklin, R., Goldstein, J., Goodall, J., Hemmerling, S., Irish, J., Lazarus, S., Loftis, D., Luther, M., McCallister, L., McGlathery, K., ... Zinnert, J. C. (2021). Anticipating and adapting to the future Impacts of climate change on the health, security and welfare of Low Elevation Coastal Zone (LECZ) communities in Southeastern USA. *Journal of Marine Science and Engineering, 9*, 1196. https://doi.org/10.3390/jmse9111196

Allison, M.,Yuill, B.,Törnqvist, T., Amelung, F.,Dixon, T. H., Erkens, G., Stuurman, R., Jones, C., Milne, G., Steckler, M., Syvitski, J., & Teatini, P. (2016). Global risks and research priorities for coastal subsidence. *Eos, 97.* https://doi.org/10.1029/2016EO055013

Ao, Z., Hu, X., Tao, S., Hu, X., Wang, G. Li, M., Wang, F., Hu, L., Liang, X., Xiao, J., Yusup, A., Qi., W., Ran, Q., Fang, J., Chang, J., Zeng, Z., Fu, Y. H., Xue, B.-L., Wang, P., ... Fang, J. (2024). A national assessment of land subsidence in China's major cities. *Science*, 384(6693), 301–306. https://www.science.org/doi/10.1126/science.adl4366

Banzhaf, E., Bulley, H. N., Inkoom, J. N., & Elze, S. (2022). Mapping open data and big data to address climate resilience of urban informal settlements in Sub-Saharan Africa. *Climate, 10*(12), 186. https://doi.org/10.3390/cli10120186

Bucci, L., Alaka, L., Hagen, A., Delgado, S., & Beven, J. (2023). *Hurricane Ian (AL09022): 23030 September 2022*. National Hurricane Center Tropical Cyclone Report. National Oceanic and Atmospheric Administration. https://www.nhc.noaa.gov/data/tcr/AL092022_Ian.pdf

Cornwall, W. (2022). Fiery invasions. *Science, 377*(6606), 568–571. https://doi.org/10.1126/science.ade2171

Cho, H.-T., Kwon, S.-H., & Han, J.-G. (2017). Estimation of environmental load of geotechnical structure using multiple regression analysis. *KSCE Journal of Civil Engineering, 21*(5), 1581–1586. https://doi.org/10.1007/s12205-017-0419-y

Cialone, M. A., Straub, J. A., Raubenheimer, B., Brown, J. A., Brodie, K. L., Elko, N., Dickhudt, P. J., Forte, M. F., DeLoach, S. R., Stockdon, H. F., & Rosati, J. D. (2023). *Large-scale community storm processes field experiment, ERDC/CHL TR-23-3*. Engineer Research and Development Center. https://doi.org/10.21079/11681/46548

Davydzenka, T., Tahmasebi, P., & Shokri, N. (2024). Unveiling the global extent of land subsidence: The sinking crisis. *Geophysical Research Letters. 51*(4), e2023GL104497. https://doi.org/10.1029/2023GL104497

Dubois, K., Dahl Larsen, M. A., Drews, M., Nilsson, E., & Rutgersson, A. (2024). Technical note: Extending sea level time series for the analysis of extremes with statistical methods and neighbouring station data, *Ocean Science, 20*(1), 21–30. https://doi.org/10.5194/os-20-21-2024

EPA. (2023). *Climate change indicators: Coastal flooding*. US Environmental Protection Agency. https://www.epa.gov/climate-indicators/climate-change-indicators-coastal-flooding#ref4

Elkington, J. (2018). Partnerships from cannibals with forks: The triple bottom line of 21st-century business. *Environmental Quality Management, 8*(1), 37–51. https://doi.org/10.1002/tqem.3310080106

Freeman, R. E., & Dmytriyez, S. (2017). Corporate social responsibility and stakeholder theory: Learning from each other. *Symphonya: Emerging Issue in Management Issue, 1*, 7–15. https://doi.org/10.4468/2017.1.02freeman.dmytriyev

Giosan, L., Syvitski, J., & Constantinescu, S. (2014). Climate change: Protect the world's deltas. *Nature, 516*, 31–33. https://doi.org/10.1038/516031a

Hall, C. (2021, April 2). Disappearing deltas: The Delta-X airborne mission investigates. Earth Data: Open Access for Open Science. https://www.earthdata.nasa.gov/learn/articles/disappearing-deltas?utm_source=eoannounce&utm_medium=email&utm_campaign=articles

Jackson, D. W. T., & Short, A. D. (Eds.), *2020 Sandy Beach Morphodynamics*, p. 817. The Netherlands, (2020)

Klotzbach, P. J., Wood, K. M., Bell, M. M., Blake, E. S., Bowen, S. G., Caron, L.-P., Colling, J. M., Gibney, E. J., Schreck, C. J., III., & Truchelut, R. E. (2022). A Hyperactive end to the Atlantic hurricane season October–November 2020. *Bulletin of the American Meteorological Society, 103*(1), E110–E128. https://doi.org/10.1175/BAMS-D-20-0312.1

Nienhuis, J. H., Heijkers, L. G. H., & Ruessink, G. (2021). Barrier breaching versus overwash depositon. Predicting the morphologic impact of storms on coastal barriers. *Journal of Geophysical Research Earth Surface, 126*(6), e2021JF006066, https://doi.org/10.1029/2021JF006066

Overeem, I. & Syvitski, J. P. M. (2009). *Dynamics and Vulnerability of Delta Systems. LOICZ Reports & Studies No. 35*, p. 54. GKSS Research Center, Geesthacht.

Pollard, J. A., Spencer, T., & Jude, S. (2018). Big data approaches for coastal flood risk assessment and emergency response. *Wires Climate Change., 9*(5), e543. https://doi.org/10.1002/wcc.543

Ranasinghe, R., Callaghan, D. P., Li, F., Wainwright, D. J., & Duong, T. M. (2023). Assessing coast-line recession for adaptation planning: Sea level rise versus storm erosion. *Scientific Reports., 13*, 8286. https://doi.org/10.1038/s41598-023-35523-8

Sarker, M. N. I., Peng, Y., Yiran, C., & Shouse, R. C. (2020). Disaster resilience through big data: Way to environmental sustainability. *International Journal of Disaster Risk Reduction, 51*, 101769. https://doi.org/10.1016/j.ijdrr.2020.101769

Senge, P. M., & Sterman, J. D. (1992). Systems thinking and organizational learning: Acting locally and thinking globally in the organization of the future. *European Journal of Operational Research., 59*(1), 137–150. https://doi.org/10.1016/0377-2217(92)90011-W

Straub, J. A., Cialone, M. A., Raubenheimer, B., Brown, J. A., Elko, N., & Brodie, K. L. (2023). The During Nearshore event Experiment (DUNEX): A collaborative coastal community experiment to address coastal resilience. *Shore & Beach, 91*(3), 23–29. https://doi.org/10.34237/1009133

Törnqvist, T. E., Jankowski, K. L., Li, Y-X., & González, J. L. (2020). Tipping points of Mississippi Delta marshes due to accelerated sea-level rise. *Science Advances, 6*(21), eaaz5512. https://doi.org/10.1126/sciadv.aaz5512

Trtanj, J., Jantarasami, L., Brunkard, J., Collier, T., Jacobs, J., Lipp, E., McLellan, S., Moore, S., Paerl, H., Ravenscroft, J. & Sengco, M. (2016). Ch. 6: Climate impacts on water-related illness. *The Impacts of Climate Change on Human Health in the United States: A Scientific Assessment*, pp. 157–188.

Turner, R. E., Kearney, M. S., & Parkinson, R. W. (2018). Sea-level rise tipping point of delta sur-vival. *Journal of Coastal Research, 34*(2), 470–474. https://doi.org/10.2112/JCOASTRES-D-17-00068.1

Turner, M. G., Calder, W. J., Cumming, G. S., Hughes, T. P., Jentsch, A., LaDeau, S. L., Lenton, T. M., Shuman, B. N., Turetsky, M. R., Ratajczak, Z., & Williams, J. W. (2020). Climate change, ecosystems and abrupt change: Science priorities. *Philosophical Transactions of the Royal Society B, 375*(1794), p. 20190105.

USGCRP. (2016). *The Impacts of Climate Change on Human Health in the United States: A Sci-entific Assessment*. Crimmins, A., J. Balbus, J. L. Gamble, C. B. Beard, J. E. Bell, D. Dodgen, R. J. Eisen, N. Fann, M. D. Hawkins, S. C. Herring, L. Jantarasami, D. M. Mills, S. Saha, M. C. Sarofim, J. Trtanj & L. Ziska (Eds.), U.S. Global Change Research Program, Washington, DC, (pp. 312). https://doi.org/10.7930/J0R49NQX

Wahl, T., Jain, S., Bender, J., Meyers, S. D., & Luther, M. E. (2015). Increasing risk of compound flooding from storm surge and rainfall for major US cities. *Nature Climate Change., 5*, 1093–1097. https://doi.org/10.1038/nclimate2736

Wilson, J., & Daugherty, P. R. (2018). Collaborative intelligence: Humans and AI are joining forces. *Harvard Business Review, 96*(4), 114–123.

Wright, L. D., Wu, W., & Morris, J. (2019a). Coastal erosion and land loss: Causes and impacts. In L. D. Wright & C. R. Nichols (Eds.), *Tomorrow's coasts: Complex and impermanent*, (pp. 137–150). Springer. https://doi.org/10.1007/978-3-319-75453-6_9

Wright L. D., Syvitski, J. P. M., & Nichols, C. R. (2019b). Complex intersections of seas, lands, rivers and people. In L. D. Wright & C. R. Nichols (Eds.), *Tomorrow's coasts: Complex and impermanent* (pp. 59–68). Springer. https://doi.org/10.1007/978-3-319-75453-6_4

Wright, L. D., & Thom, B. G. (2023). Coastal morphodynamics and climate change: A review of recent advances. *Journal of Marine Science and Engineering, 11*(10), 1997. https://doi.org/10.3390/jmse11101997

World Health Organization. (2003). The world health report 2003: Shaping the future. *World Health Organization*. https://iris.who.int/bitstream/handle/10665/42789/9241562439.pdf

Vulnerability: The Opposite of Resilience

4

Vulnerability defines the inability of people to resist and adapt to climate hazards. It relates to the state of being exposed to the possibility of being harmed by climate change hazards and depends on factors including location, construction, contents, and economic value. It could be expressed as (Balica & Wright, 2010):

$$\text{Vulnerability} = \text{Exposure} + \text{Susceptibility} - \text{Resilience}. \qquad (4.1)$$

Thus, people living in floodplains, along the coast, or in areas prone to severe storms are more vulnerable. Similarly, those living in poverty may be less able to prepare for or to respond to the effects of climate change. Numerous authors (Derakhshan et al., 2022; Emrich & Cutter, 2011; and Habets et al., 2023) have provided insights on conducting viable vulnerability assessments. They have also discovered that different populations face different levels of vulnerability and risk. Assessments support the development of policies focused on addressing vulnerability and risk and disaster outcomes. FEMA defines a disaster as "a non-routine event that exceeds the capacity of the affected area to respond to it in such a way as to save lives; to preserve property; and to maintain the social, ecological, economic, and political stability of the affected region." One way to reduce vulnerabilities is to build capacities (Fig. 4.1) over time to increase the ability of a community to resist and recover from a disaster. Absorption could include restoring natural features that help protect the coast from waves and sea level rise. For example, engineering projects could restore dunes, oyster and submerged aquatic vegetation beds, coral reefs, salt marshes, and mangrove forests. Adaptation could include educational programs that raise awareness about coastal threats. Transformative approaches include *building* infrastructure that can withstand floods, droughts, and other extreme weather conditions and implementing policies and programs that protect and restore natural wetlands.

© The Author(s), under exclusive license to Springer Nature Switzerland AG 2025
C. Reid Nichols et al., *Integrated Coastal Resilience*, Synthesis Lectures on Ocean
Systems Engineering, https://doi.org/10.1007/978-3-031-68153-0_4

Fig. 4.1 Resilient capacities should be transformative, absorptive, and adaptive. Transformative activities address underlying failures that cause or increase risks. Absorption involves taking corrective actions such as restoring wetlands that act as a buffer to storms and flooding. Adaptation involves making appropriate changes in order to improve resilience such as enabling people to move to higher ground

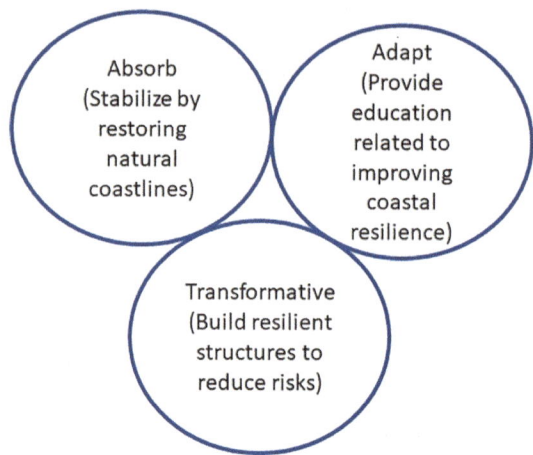

The effects of climate change can threaten routine tasks, a business' critical success factors, engineering operations, and a project's stakeholders. Many governments and their agencies have defined costs ranging from adaptation to innovation that are associated with climate change (e.g., Lipier, 2021). One should use available tools such as observing systems or weather stations to collect information on how frequently an indicator such as nuisance floods have occurred in the region of concern. The data should help to determine whether or not climate change or other stressors are likely to increase the frequency or severity of natural hazards. One might follow a classic risk management approach to cope with the effects of climate change (PMI, 2021). Implementing this approach to improve coastal resilience involves working with stakeholders to identify the most important climate change risks or hazards in advance, prioritizing them, and developing the appropriate responses for each effect. There are three general steps to undertake. First, based on data, identify the hazards that are appropriate for the area or region. Second, prioritize the threats by considering their probability and impact. Finally, determine viable responses to each risk or hazard. Risk analysis will improve your resilience and reduce your vulnerability. Identified risks may or may not occur, but identification is a first step toward preparation and responding to changing conditions. Organizations from small business to national governments are considering the impact of climate change on their operations.

As indicated in earlier sections of this book, one might use sources such as historic data, real-time data from sensor networks, imagery, and model output to identify risks, which might come in the form of precipitation[1], wind, waves, current, and other external forces. For example, water level gage networks document how rising sea levels inundate low-lying wetlands and dry land, erode shorelines, contribute to coastal flooding, and increase the flow of salt water into estuaries and nearby groundwater aquifers. According

---

[1] Nowadays, extreme rains and floods in the UAE, Oman, and Bahrain seem to have become a norm (e.g., Jonathan & Raju, 2017; Zachariah et al., 2024).

to PMI (2021), a risk management plan would be prepared with components such as a risk register, risk breakdown structure, and risk analysis. The risk register would include an itemized listing of the most important climate change effects and your vulnerability or susceptibility to physical injury, harm, damage, or economic loss. For example, the risk register might consider the probability and impact of a tropical cyclone or a heatwave. One might consider reductions in dissolved oxygen in warm sea water, which is the main cause of coral bleaching (Cornwall, 2023), but might have other impacts on a specific area, such as fish kills or rotting algal mats on the shoreline. One might deal with the many climate change factors by dividing the risks into hierarchical categories or into a Risk Breakdown Structure (RBS). This RBS might define total risk exposure for a particular phenomenon, where descending levels represent increasingly detailed definitions for associated parameters and data. Engineers or city planners would be able to divide risks into categories that will help associate risks with specific project aspects and stakeholders.

4.1   Risk Management

The identification of risks with your stakeholders is a preliminary step to assessing your vulnerability. Accounting for risks will have different levels of importance for your stakeholders. Therefore, developing a risk management plan requires positive stakeholder engagement. Prioritizing risks ensures that stakeholders can recognize the importance placed on their areas of concern, which ensures that you are benefiting stakeholders (Schaltegger et al., 2019). In communication with stakeholders determine the probability and impact of each risk as illustrated in Eq. 4.1.

$$\text{Risk} = \text{Probability} \times \text{Impact} \qquad (4.2)$$

Each of the risk factors should be prioritized using a scale such as low, medium, or high or possibly 1–10. PMI (2021) provides various ways to present risks and one example is provided in Table 4.1. In this case climate change risks are listed in the first column, followed by the probability, impact rating, and a score. One might prioritize each risk based on Eq. 4.1. Impact ratings could be based on factors such as a severity score, time, cost, money or a combination of factors. After the initial ranking, an overall prioritization could be determined to benefit stakeholders. The table provides the reader with an appreciation for which risks need to be acted on to improve resilience. After a risk event such as flooding occurs, it becomes an issue. Issues require a response, usually from engineering project managers, environmental scientists, community planners or property owners. Effective use of capabilities such as an ocean observing systems or a Physical Oceanographic Real-Time System (PORTS®; see https://tidesandcurrents.noaa.gov/ports.html) enables planners to identify triggers for phenomena such as extreme water level fluctuations. Identifying risk triggers is an important component of risk analysis

Table 4.1 Notional risk register

Risk	Probability	Impact	Score	Trigger	Response
Extreme winds	0.30	3	0.9	Winds > 25 m/s	Shutter windows
Flooding	0.26	3	0.68	2 ft > high tide	Close roads
High waves	O.33	2	0.66	Wave heights > 2 m	Small craft warnings
Intense rain	0.25	2	0.5	Rainfall > 100 mm/24 hrs	Get to high ground
River discharge	0.3	1	0.3	Turbidity > 100 NTU	Survey bridges
Extreme heat	0.01	3	0.03	Temp. > 32.2 °C for 3 days	Burn ban

An impact scale was used where 4 = catastrophic, 3 = critical; 2 = marginal; and 1 = negligible

and supports operational organizations such as NOAA and the USGS to design effective monitoring systems that allow communities to mitigate threats.

A risk register supports planning by those involved in resilience, which includes stakeholders with varying backgrounds and concerns. The risk register helps to identify issues and responses as part of the planning process. Once a risk such as sea level rise is defined, then adaptation such as elevating houses, constructing sea walls or relocating towns can be considered. The risk register is the foundation to improve coastal resilience. Science based triggers allow the right people to know when a risk has occurred and the response that is required. According to PMI (2021) the four general responses possible for a risk such as flooding are: (1) avoiding the risk by elevating a structure; (2) transferring the risk to a third party (e.g., by buying flood insurance); (3) mitigating the risk by reducing the probability or impact by moving a structure away from the shoreline; and (4) accepting the fact that nuisance floods are occurring during spring tides.

The phenomenon of climate-related displacement is being studied as extreme weather events become more common and more intense. Disasters of all types displace people and those who are displaced become vulnerable to abuse, exploitation, poverty, and violence. Displacement exacerbates the impact of shocks, such as flooding in South Asia after Cyclone Amphan (2021) and more frequent droughts in sub-Saharan Africa. Climate refugees reduce resilience and increase vulnerability. Bellizzi et al. (2023) have suggested that this rising crisis requires early warning systems and preparedness to prevent and address situations of vulnerability. Improving resilience helps to mitigate the strong relationships among climate change, migration, and conflict that was reported by Reuveny (2007). Finding solutions to displacement will be a process that requires real collaboration among stakeholders. Solutions such as improving resilience may be more realistic than options related to return, local integration, and relocation. Immediate solutions such as local integration are fraught with complex legal, economic, social and cultural challenges.

4.2 Responding to Threats

Resilience requires planning for and practicing the response to threats. Organizations such as FEMA provide planning guidance for how a community can plan a viable response for a potential crisis, determining required capabilities and establishing a framework for roles and responsibilities (see https://www.fema.gov/emergency-managers/national-prepar edness/plan). Engineers, environmental scientists, and the community should collaborate to determine their information requirements. Responding to issues such as a flood could result in large volumes of data from sensor networks, satellites, social media servers, cell phones and other multimedia devices. In some cases, government organizations and universities provide tools and technical expertise to communities to better understand local problems and advocate for improved environmental health. Further, communities with environmental concerns may collaborate with nonprofit organizations such as TNC and academic researchers. Citizen scientists can improve resilience by helping communities to characterize environmental problems and collect quality data. Crowd-sourced information exchanged in social media during disasters such as volcanic eruptions can be used by first responders for detecting events and even clarifying event locations (Earle et al., 2012; Wendel, 2015).

Data are a central component for resilience and must be processed at high speeds. Paul Seaton, Director for Hydrography and Coastal Resilience for Asia Pacific at Fugro put this notion into context for coastal resilience management (Guhu, 2023). Information that is collected on environmental factors that impact operations must be managed. Collaborators in coastal resilience include engineers, scientists, community leaders, and first responders. They need to address data preparedness in order to support decision makers who have to respond to climate change effectively. This response may involve the deployment of sensors, running models, and managing large volumes of disparate data. Banzhaf et al. (2022) discussed the importance of using remote sensing and GIS techniques to understand spatial and temporal dynamics and using the resultant geospatial information to enhance the resilience of urban areas in Africa. Rogeiro et al. (2018) discussed the use of high-performance computing for coastal forecasting, monitoring networks, and the storage of data related to coastal resilience. Collaboration and the practice of aggregating computing power in a way that delivers higher performance computing will continue to be a major pillar in the quest to improve coastal resilience.

Data and models support the successful installation of living shorelines that have been demonstrated as a viable means to protect property and coastal resources. Smith et al. (2020) investigated numerous shoreline protection projects in North America, Asia, and Europe that utilized physical living shoreline interventions. Organizations such as TNC have teamed with government and university scientists and communities to promote the use of vegetation and reef building materials such as oysters to stabilize estuarine coasts, bays, or tributaries. The Boyd Living Shoreline is an example large-scale living shoreline project that used vegetation native to South Carolina and hard substrates to stabilize the shoreline (O'Hara, 2022). Collaborative research has resulted in the development of the Shoreline Decision Support Tool, which provides guidance on shoreline erosion control strategies along barrier island coasts. The tool is available online and may be accessed through lead organizations such as the Virginia Institute of Marine Science (see https://cmap2.vims.edu/LivingShoreline/DecisionSupportTool/index.html). The use of natural vegetation such as seaweeds, mangroves, and marsh grasses along with reef building materials to restore oyster beds not only connects the land to coastal waters, but stabilizes shorelines by reducing erosion, and providing valuable habitat that improves coastal resilience.

Coastal data are generally viewed as a public good[2] and made available through data centers or international programs such as GOOS and national programs such as US IOOS and EuroGOOS. Observed data in repositories such as the International Comprehensive Ocean-Atmosphere Data Set (ICOADS) offer surface marine data spanning the timeframe from 1662 to the present. ICOADS data are incorporated into numerous marine science products from charts to forecasts (see https://icoads.noaa.gov/). The WMO is developing a WIGOS Data Quality Monitoring System (see https://wdqms.wmo.int/), which will be hosted by the European Centre for Medium-Range Weather Forecasts to monitor the performance of all WIGOS observing components. Whereas some data are public, other data may be private and available only to select parties. In addition to public organizations, private companies may use these big data resources to improve operational efficiency. For example, shipping and rail companies perform analyses to understand and address climate change impacts. Global data in remote areas are also important, e.g., to ensure safety at sea as more vessels travel across hazardous locations like the Arctic Ocean (Müller et al., 2023). Similarly, observations from navigators and other users of maritime products may be incorporated into the planning for the update of information that is provided by operational organizations. Identifying errors in navigation products owing to the impact of climate change could be determined from observations by citizen scientists. Companies such as ProRail in Netherlands and Network Rail in the UK perform climate vulnerability and risk analysis to support design, construction, and maintenance of the rail system. The availability of quality controlled coastal data are critical for governments, scientists, industry, and communities.

References

Balica, S., & Wright, N. G. (2010). Reducing the complexity of the flood vulnerability index. *Environmental Hazards, 9*(4), 321–339. https://doi.org/10.3763/ehaz.2010.0043

Banzhaf, E., Bulley, H. N., Inkoom, J. N., & Elze, S. (2022). Mapping open data and big data to address climate resilience of urban informal settlements in Sub-Saharan Africa. *Climate, 10*(12), 186. https://doi.org/10.3390/cli10120186

---

[2] The scope of public goods can be local, national, or global.

Bellizzi, S, Popescu, C., Panu Napodano, C. M., Fiamma, M., & Cegolon, L. (2023). Global health, climate change and migration: The need for recognition of "climate refugees". *Journal of Global Health.* https://jogh.org/2023/jogh-13-03011

Cornwall, W. (2023). Breathless oceans. *Science, 379*(6631), 429–433. https://doi.org/10.1126/science.adg9640

Derakhshan, S., Emrich, C. T., & Cutter, S. L. (2022). Degree and direction of overlap between social vulnerability and community resilience measurements. *PLoS ONE, 17*(10), e0275975. https://doi.org/10.1371/journal.pone.0275975

Earle, P. S., Bowden, D. C., & Guy, M. R. (2012). Twitter earthquake detection: Earthquake monitoring in a social world. *Annals of Geophysics, 54*(6), 708–715. https://doi.org/10.4401/ag-5364

Emrich, C. T., & Cutter, S. L. (2011). Social vulnerability to climate-sensitive hazards in the Southern United States. *Journal of Weather, Climate, and Society, 3*(3), 193–208. https://doi.org/10.1175/2011WCAS1092.1

Guhu. A. (2023). *Understanding coastal resilience and geo-data: An interview with Paul Seaton.* Earth.Org. https://earth.org/interview/understanding-coastal-resilience-and-geo-data-an-interview-with-paul-seaton/

Habets, M., Jackson, S. L., Baker, S. L., Huang, Q., Blackwood, L., Kemp, E. M., & Cutter, S. L. (2023). Evaluating the quality of state hazard mitigation plans based on hazard identification, risk, and vulnerability assessments. *Journal of Homeland Security and Emergency Management.* https://doi.org/10.1515/jhsem-2022-0060

Jonathan, K. H., & Raju, P. S. (2017). Analysis of rainfall pattern and temperature variation in three regions of Sultanate of Oman. *International Journal of Civil Engineering and Technology., 8*(2), 173–181.

Lipier, E. (2021). *Climate change adaptation: Department of Commerce.* CRS Report R46743, Congressional Research Service. https://crsreports.congress.gov/product/pdf/R/R46743

Müller, M., Knol-Kauffman, M., Jeuring, J., & Palerme, C. (2023). Arctic shipping trends during hazardous weather and sea-ice conditions and the Polar Code's effectiveness. *npj Ocean Sustain, 2*, 12. https://doi.org/10.1038/s44183-023-00021-x

O'Hara, D. E. (2022). *The importance of wetlands and creating policy to protect wetlands in Georgetown County Goal 14.* Goal 14: Life Below Water. https://digitalcommons.coastal.edu/goal-14-life-below-water/1

PMI. (2021). *A Guide to the project management body of knowledge (PMBOK® guide)* (7th ed.), Project Management Institute.

Reuveny, R. (2007). Climate change-induced migration and violent conflict. *Political Geography, 26*(6), 656–673. https://doi.org/10.1016/j.polgeo.2007.05.001

Rogeiroa, J., Rodrigues, M., Azevedo, A., Oliveira, A., Martins, J. P., David, M., Dias, P., & Gomes, J. (2018). Running high resolution coastal models in forecast systems: Moving from workstation and HPC cluster to cloud resources. *Advances in Engineering Software, 117*, 70–79. https://doi.org/10.1016/j.advengsoft.2017.04.002

Schaltegger, S., Hörisch, J., & Freeman, R. E. (2019). Business cases for sustainability: A stakeholder theory perspective. *Organization & Environment, 32*(3), 191–212. https://doi.org/10.1177/10860266177228

Smith, C. S., Rudd, M. E., Gittman, R. K., Melvin, E. C., Patterson, V. S., Renzi, J. J., Wellman, E. H., & Silliman, B. R. (2020). Coming to terms with living shorelines: A scoping review of novel restoration strategies for shoreline protection. *Frontiers in Marine Science, 7*, 433. https://doi.org/10.3389/fmars.2020.00434

Wendel, J. (2015). Internet users act as earthquake trackers. *Eos*, *96*, https://doi.org/10.1029/2015EO
    025457

Zachariah, M., Kimutai, J., Barnes, C., Gryspeerdt, E., Seneviratne, S. I., Almazroui, M., Vautard,
    R., Zhang, X., Pinto, I., Vahlberg, M., Sengupta, S., Saeed, F., Otto, F. E. L., Clarke, B. Philip,
    S., Lohmann, U., Wernli, H., Mistry, M., El Hajj, R., Singh, R., & Arrighi, J. (2024). *Heavy pre-
    cipitation hitting vulnerable communities in the UAE and Oman becoming an increasing threat as
    the climate warms*. Report, Imperial College London. https://spiral.imperial.ac.uk/handle/10044/
    1/110910

Effective Management

5

Whereas there are many leadership and management styles, effective managers have the skills and confidence of getting things done efficiently and effectively with and through other people and organizations. Some of the key skill sets to manage a resilience project are described in Fig. 5.1. Substantial knowledge and technical skills are also critical and help to build trust among partners. According to PMI (2021), effective management involves cost control, engaging stakeholders, identifying risks, team building, quality control, and scheduling, all to achieve organizational goals. Toward this, organizations such as TNC have used organizational assets that include a network of practitioners around the world to support communities with hazard mitigation and climate adaptation planning.

Climate change is a major challenge that can be best approached through transdisciplinary methods, e.g., among engineers, environmental scientists, and the community. Solutions are generated through the collaboration of academic and non-academic players who have stakes in the problem. The process begins through planning and developing strategies to cope with climate change factors. An initial task will involve identifying likely risks, and working together to understand and prioritize the risks based on their occurrence and severity. A resulting plan of action can be developed to reduce and mitigate those risks. This risk planning process is essential to prepare, resist, recover, and adapt to climatic disturbances. Solutions will help the community to recover from the impacts of natural hazards such as extreme weather. Collectively, and possibly with the support of an intermediary, organizations might assess progress towards minimizing risks.

There are a variety of environmental factors (e.g., temperature and precipitation) that determine the frequency of climatic effects that impact government, industry, and community operations or activities. Climate change can, for example, affect government spending programs, industry through supply chain interruptions, and disrupt the natural, economic

C. Reid Nichols et al., *Integrated Coastal Resilience*, Synthesis Lectures on Ocean Systems Engineering, https://doi.org/10.1007/978-3-031-68153-0_5

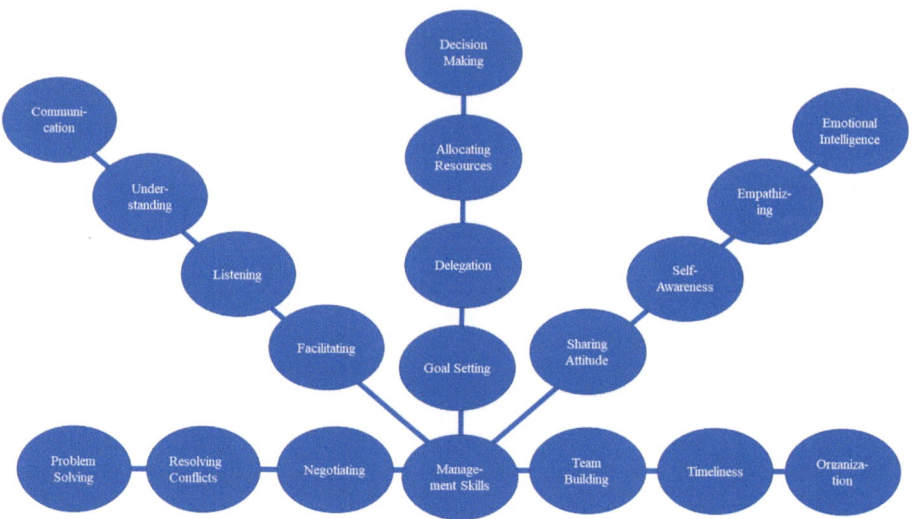

Fig. 5.1 Example skills that support the effective management of collaborative projects. Many of these skills were included in a work by O'leary et al. (2012) that described the views Senior Executives from the U.S. with experience supervising collaborative projects. Partners in R&D may be working on multiple teams

and social systems that we all depend on. Identifying the common risks in a locale can help get the risk management process started. Some examples include damage caused by natural disasters such as floods or wildfires. Identifying risk factors with the help of stakeholders and how they impact your system is essential to improving coastal resilience.

Once risk factors are determined, they need to be prioritized by the potential frequency that they might occur and the potential impact that they could have. Aven (2016) described foundations for risk assessment and management. Developing risk management techniques and strategies are essential to establishing viable coastal resilience projects. Understanding which risks have the most detrimental impact and are most likely to occur narrows down the top priorities for an area's resilience projects.

The planning and execution of coastal resilience projects is often beyond the scope of a single organization and generally involves meeting diverse requirements that benefit a variety of stakeholders. Climate change issues often require many centers of decision-making authority to cover the full range of governance tasks (Morrison et al., 2019). Management is complicated owing to variations between federal and state jurisdictions, differing definitions of coastal boundaries and varying perspectives on climate change impacts (Harvey & Thom, 2024). Projects that improve coastal resilience often involve partners from government, universities, and industry. Stadtler and Karakulak (2020) discussed collaboration dynamics and benefits and pitfalls to using broker organizations to facilitate collaboration among diverse partners. Nichols and Wright (2020) highlighted several collaborative coastal research projects that were facilitated by the Southeastern Universities Research Association[1] and aimed at improving coastal resilience. Partners were from government, universities, and industry and some technologies were transferred to NOAA. From a systems perspective, broker organizations not only facilitate the partnering process, but they ensure effective after-action review and learning.

Coastal resilience projects are complex and are often unique to coastal types (e.g., barrier islands, cuspate forelands, deltas, drumlins, estuaries, mangroves, glacial moraines, volcanic, wave cut cliffs, etc.). In each of these coast types, the response to climate changes may be different. For example, temperature characteristics of a region are influenced by natural factors such as latitude, elevation, and proximity to ocean currents. Low-lying mangrove areas will flood more often than areas with wave cut cliffs. Similarly, some regions will receive more rainfall while others are exposed to droughts. Held and Soden (2006) described this phenomenon as wet regions get wetter and dry regions drier. Governments, engineers, and communities must consider location, geography, geology, meteorology, and oceanography to understand climate change's effects in a particular location.

Organizations have differing values, beliefs, assumptions, and norms that guide their activities. Harvey and Thom (2024) highlighted the importance of federal governments to support coastal resilience type programs at the national level. Local governance takes responsibility for a variety of services (e.g., housing, emergency medicine, transportation, and public works) that may be impacted by local climatic change effects (e.g., drought, fluvial flooding, wildfires). University researchers may have expertise in these areas and industry can transition or implement research results that may improve coastal resilience. Dunlop et al. (2014) highlighted that universities also provide valuable resources during natural disasters, which if leveraged correctly can improve a community's resilience. Research results are also transitioned to operations through industry support. The private sector provides skilled services in form of technical manpower or in-kind donations of goods or services to improve coastal resilience. For these reasons, a strong partnership that includes government, university, and industry collaborators has the highest probability of making a positive impact to increased coastal resilience.

Successful project completion requires navigating challenges that result from partners who are coming from multiple organizations. Organizational structure and leadership will impact how coastal resilience objectives are accomplished. Pisano and Verganti (2008) illustrated how collaborative projects could be managed through hierarchical organizations such as government agencies or by flat organizations such as intermediaries or brokers. They detailed levels of participation ranging from closed to open projects. For example, closed systems such as the Chesapeake Research Consortium (see https://chesapeake.org/) would include primary partners from Johns Hopkins University, University System

---

[1] The Southeastern Universities Research Association is a non-profit association of universities from the United States and Canada.

Table 5.1 Modes of collaboration in accordance with low or high participation in governance. Collaboration may occur across the spectrum of participation from closed to open organizations. These four general modes of collaboration were described and illustrated by Pisano and Verganti (2008)

Innovation Mall—An organization solicits proposals. The organization chooses the solution that it likes the best *Example*: The interagency National Coastal Resilience Fund has selected projects that contribute to the restoration or enhancement of natural buffers such as coastal marshes and wetlands, dune and beach systems, oyster and coral reefs, mangroves, maritime forests, coastal rivers, and barrier islands (see https://www.nfwf.org/pro grams/national-coastal-resilience-fund)	**Innovation Community**—A network where anybody can propose problems, offer solutions, and decide which solutions to use *Example*: Resilient Cities Network (R-Cities) which was pioneered by the Rockefeller Foundation. R-Cities helped to articulate and develop innovative ideas by bringing together global knowledge, practice, partnerships, and funding to empower participating cities (see https://resilientcitiesnetwork.org)	Open	Participation
Elite Circle—A select group of participants chosen by an organization that also defines the problem and picks the solutions *Example*: TNC developed the California Coastal Resilience Network to share knowledge and solutions aimed at mitigating threats from climate induced impacts to California's coastal habitats. Individuals or organizations can become Network members by joining the CA Coastal Resilience Network (see https://coa stalresilience.org/project/california-coastal-resilience-network/)	**Consortium**—A private group of participants that jointly select problems, decide how to conduct work, and choose solutions *Example*: Chesapeake Research Consortium (CRC) is an association of seven research and education institutions around the Chesapeake Bay region. With funding from organizations such as the interagency Chesapeake Bay Program, CRC focuses research on Chesapeake Bay restoration (see https://chesapeake.org/)	Closed	
Hierarchical	Flat		
Governance			

of Maryland, Smithsonian Institution, Virginia Institute of Marine Science, Old Dominion University, or Pennsylvania State University. *Open systems such as TNC's Coastal Resilience (see* https://coastalresilience.org/) are communal where shared values and purpose guide participation. Resilience projects are fit into the Pisano and Virganti matrix that is provided in Table 5.1 to illustrate this conceptual framework.

Partnership projects, while complicated to manage, improve the potential for success through shared knowledge, expertise, and resources. Collaboration is a factor in the success of environmental projects and should be considered to improve coastal resilience (e.g., living shoreline projects). Managers should facilitate the sharing of partner capabilities, resources, and knowledge to share risk, increase innovation and take advantage of disparate knowledge. For these reasons, broker organizations have shown to influence success for public–private partnerships (Stadtler & Karakulak, 2020; Stadtler & Probst, 2012). A trusted third-party broker can facilitate partnerships among physical and social scientists, engineers, and the community while supporting implementation of research results. Effective collaborators share resources to achieve objectives quickly. Nichols and Wright (2020) described the management of partnerships that contributed to the development of the Coastal and Ocean Modeling Testbed for NOAA. Effective management refers to how managers from multiple organizations achieve their targets with team resources. Effective partnering can result in a significantly higher level of quality on a project and can significantly increase the probability of timely and on-budget completion of the project.

Systems thinking provides a foundation for project planning and implementation. To avoid unintended consequences,[2] the system within which a resilience project will be performed must be understood as well as the implications for local and regional stakeholders (Senge, 1990). Management researchers have illustrated that collaboration among diverse and varied stakeholders mitigated negative outcomes and was essential to the success of projects (Freeman, 1984). Collaboration among government, university, and industry entities helps to efficiently use scarce resources while ensuring that project effects or consequences are understood by all the stakeholders, especially the local community. Collaboration should be managed to ensure that coastal resilience projects produce solutions to problems that meet the described needs of the collaborators and results are implementable.

The notion of engaging broad stakeholders in problem solving as co-designers can be traced to Scandanavia. Bødker et al. (1993) highlighted the development of cooperative design to empower users through participation in product development, especially for computer applications. Reed (2008) described the benefits of stakeholder engagement for decision making. Participation by stakeholders in a project expands the knowledge and resources that are available for solutions, fosters realistic expectations, reduces resistance to change, and ensures the development of effective solutions. Prahalad and Ramaswamy (2000) coined the term "co-creation" as they described the benefits for a company to harness the power of customers and other stakeholders. Engaging stakeholders effectively is a key project management task and includes activities such as holding stakeholder workshops that facilitate collective creativity. Social and behavior change communication might be applied to transfer knowledge and positively influence attitudes and social norms among individuals, institutions and communities. The Global Ocean Observing

---

[2] Unintended consequences can be either positive or negative.

System has embarked on such a program to ensure the development of effective observing systems. The examplar projects listed in Table 5.2 have been established to include stakeholders in the development of best practices and tools for observing system design (GOOS, 2022). To improve coastal resilience, the application of co-design would encourage participation of the general community with scientists and engineers who are working on resilience projects.

Table 5.2 Co-design as applied by the GOOS program includes outreach to government, university, and industry partners. The below exemplar projects are being implemented as use-cases where efforts are underway to understand requirements for users of ocean information

Exemplar	Potential end-users	Leads
Tropical cyclones in Caribbean, Indian Ocean, and North Pacific Ocean and marginal seas	National forecasting systems	Multiple organizations such as the Korean Institute of Ocean Science and Technology; National Taiwan University; NOAA; Ocean University of China; Rutgers; and University of the Ryukyus are advancing an Earth System Modeling approach where components of a *climate model* simulate the atmosphere, the ocean, sea, ice, the land surface and the vegetation on land and the biogeochemistry of the ocean
Marine life	Local coastal managers, planners, industries, and communities	Researchers from University of South Florida and Universiti Sains Malaysia are leading an effort to develop tools and management frameworks that build resilience, recognize thresholds and avoid ecological tipping points (Muller-Karger et al., 2023)
The ocean carbon cycle	National policy makers	Researchers affiliated with the proposed The North Atlantic Carbon Observatory (NACO) at the Ocean Frontiers Institute of Dalhousie University. NACO will link intergovernmental agencies, industry, and research to deliver ocean carbon measurements at basin and climate-relevant scales
Boundary currents	Marine transportation, fisheries, forecasters, marine spill response, mining and oil and gas industry	Organizations such as the South African Weather Service and NOAA are leading an effort focused on observation and monitoring of the Agulhas Current (Morris et al., 2022)
Marine heatwaves	Aquaculture, fisheries, national and local resource managers	Researchers from the Sorbonne Université, Research Consortium for the Gulf of Mexico (CIGoM), and their stakeholders are working on better monitoring and modeling systems in regions that include the Mediterranean Sea, Caribbean Sea, and West Africa

(continued)

Table 5.2 (continued)

Exemplar	Potential end-users	Leads
Storm surge	Local coastal managers, planners, industries, and communities	Researchers from Centro Euro-Mediterraneo sui Cambiamenti Climatici (CMCC) and their stakeholders are working to increase the accuracy in forecasting storm surge for vulnerable communities

These UN Decade of Ocean Science for Sustainable Development (2021–2030) "Exemplar Projects" aim to enhance ocean observing systems in areas related to tropical cyclones, marine life, ocean carbon, boundary current, marine heatwaves, and storm surge (see https://www.une sco.org/en/decades/ocean-decade)

References

Aven, T. (2016). Risk assessment and risk management: Review of recent advances on their foundation. *European Journal of Operational Research, 253*(1), 1–13. https://doi.org/10.1016/j.ejor.2015.12.023

Bødker, S., Kyng, M. & Grønbæk, K. (1993). Cooperative design: Techniques and experiences from the Scandinavian scene. In D. Schuler & A. Namioka (Eds.), *Participatory design: Principles and practices*, CRC Press. https://doi.org/10.1201/9780203744338-8

Dunlop, A. L., Logue, K. M., & Isakov, A. P. (2014). The engagement of academic institutions in community disaster response: A comparative analysis. *Public Health Reports, 129*(6), 87–95. https://doi.org/10.1177/00333549141296S4

GOOS. (2022). *Exemplar explainer document*. The Global Ocean Observing System. https://www.goosocean.org/

Freeman, R. E. (1984). *Strategic management: A stakeholder approach*. Pitman Publishing Inc.

Harvey, N. & Thom, B. (2024). Coastal governance in federated countries. In D. Baird & M. Elliott (Eds.), *Treatise on estuarine and coastal science* (2nd ed.), (pp. 307–326). Academic Press. https://doi.org/10.1016/B978-0-323-90798-9.00098-6

Held, I. M., & Soden, B. J. (2006). Robust responses of the hydrological cycle to global warming. *Journal of Climate, 19*(21), 5686–5699. https://doi.org/10.1175/JCLI3990.1

Morrison, T. H., Adger, W. N., Brown, K., Lemos, M. C., Huitema, D., Phelps, J., Evans, L., Cohen, P., Song, A. M., Turner, R., Quinn, T., & Hughes, T. P. (2019). The black box of power in polycentric environmental governance. *Global Environmental Change, 57*, 101934. https://doi.org/10.1016/j.gloenvcha.2019.101934

Morris, T., Rudnick, D., Sprintall, J., Hermes, J., Goni, G. J., Parks, J., Bringas, F., Heslop, E., and the numerous contributors to the OCG-12 Boundary Current Workshop and OceanGliders BOON Project. (2022). Monitoring boundary currents using ocean observing infrastructure. *Frontiers in Ocean Observing: Documenting Ecosystems, Understanding Environmental Changes, Forecasting Hazards. 34*(4), 16–17. https://doi.org/10.5670/oceanog.2021.supplement.02-07

Muller-Karger, F. E., Canonico, G., Aguilar, C. B., Bax, N. J., Appeltans, W., Yarincik, K., Leopardas, V., Sousa-Pinto, I., Nakaoka, M., Aikappu, A., Giddens, J., Heslop, E., Montes, E., & Duffy, J. E. (2023). Marine Life 2030: Building global knowledge of marine life for local action in the ocean decade, *ICES Journal of Marine Science, 80*(2), 355–357. https://doi.org/10.1093/icesjms/fsac084

Nichols, C. R., & Wright, L. D. (2020). The evolution and outcomes of a collaborative testbed for predicting coastal threats. *Journal of Marine Science and Engineering, 8*(8), 612. https://doi.org/10.3390/jmse8080612

O'leary, R., Choi, Y., & Gerard, C. M. (2012). The skill set of the successful collaborator. *Public Administration Review., 72*(1), 570–583. https://doi.org/10.1111/j.1540-6210.2012.02667.x

PMI. (2021). *A Guide to the project management body of knowledge (PMBOK® guide)* (7th ed.). Project Management Institute.

Pisano, G. P., & Verganti, R. (2008). Which type of collaboration is right for you? *Harvard Business Review, 88*(12), 78–86.

Prahalad, C. K., & Ramaswamy, V. (2000). Co-opting customer confidence. *Harvard Business Review, 78*(1), 79–87.

Senge, P. M. (1990). *The fifth discipline: The art and practice of the learning organization.* Doubleday.

Stadtler, L., & Karakulak, Ö. (2020). Broker organizations to facilitate cross-sector collaboration: At the crossroad of strengthening and weakening effects. *Public Administration Review, 80*(3), 360–380. https://doi.org/10.1111/puar.13174

Stadtler, L., & Probst, G. (2012). How broker organizations can facilitate public-private partnerships for development. *European Management Journal, 30*(1), 32–46. https://doi.org/10.1016/j.emj.2011.10.002

Reed, M. S. (2008). Stakeholder participation for environmental management: A literature review. *Biological Conservation, 141*(10), 2417–2431. https://doi.org/10.1016/j.biocon.2008.07.014

Closing Thoughts on Future Adaptations

For a community to be resilient and "bounce back" after hazardous events such as tropical cyclones, coastal storms, and flooding, an integrated approach must be put into practice. Decision makers need to develop plans based on disparate forms of data. Quantifiable and qualitative data collections must be actionable and synthesized to support the development of viable plans. Rather than simply reacting to impacts, data and information must be collected to inform managers, engineers, environmental scientists, and citizen scientists. These data are from historical archives, local knowledge, weather stations and networked observatories, satellites, aircraft, vessels, and forecasters. The data may be used extensively to drive engineering designs which improve coastal resilience. Scientists and engineers can apply these data to ensure that a community is informed and prepared to rebound quickly from weather and climate-related events, including adapting to sea level rise. Expanding the availability of fully integrated and operational networks such as NOAA Mesonet (see https://nationalmesonet.us/) will help to improve environmental characterization and weather forecasting. In meteorology, a mesonet is a network of automated weather and environmental monitoring stations designed to observe mesoscale meteorological phenomena and microclimates. They are particularly useful for agriculture. Networked sensors provide an opportunity for engineers and scientists to build innovative decision aid products or dashboards that can support routine tasks from bridge maintenance to navigation to assessing wetland status to fishery trends to emergency response. Other example successes with networks include Long-term Ecological Research (see https://lternet.edu/), NDBC Buoys (see https://www.ndbc.noaa.gov/), PORTS® (see https://tidesandcurrents.noaa.gov/ports.html), and WIGOS (see https://community.wmo.int/en/activity-areas/WIGOS).

C. Reid Nichols et al., *Integrated Coastal Resilience*, Synthesis Lectures on Ocean Systems Engineering, https://doi.org/10.1007/978-3-031-68153-0_6

In some cases, collaborative projects that include government, university, and industry researchers benefit from an intermediary. Tasks of a broker organization ensure effective sharing of resources and the transition of research results to operations or commercialization. One element that broker organizations should accomplish involves effecting positive stakeholder engagement. Accomplishments with stakeholders range from collecting and understanding requirements to creative design. Improving the resilience of a coastal region requires a team approach. Impacting factors span the engineering and the physical and social sciences. Integrated approaches to resilience are just one factor to improving the ability of a community or city to rebound more quickly while reducing negative human health, environmental, and economic impacts.

Engineered structures are increasingly dominating coastal landscapes as urbanization increases along with climate changes that are accelerating coastal threats. These structures (e.g., artificial dunes, seawalls, revetments, breakwaters, jetties, groins, and bulkheads) are designed and approved by professional engineers. The USACE has played a leading role in the design and construction of many of these protective solutions. A prominent example is the flood protection system surrounding the city of New Orleans that was developed following the Hurricane Katrina catastrophe in 2005. However, some less well planned "grey infrastructure" implemented by inexperienced or untrained regional planners or engineers have often exacerbated local erosion. For example, shoreline perpendicular structures such as groins interrupt the longshore transport of sand along a beach. Sand is accreted on the updrift side of the groin causing erosion to occur on the downdrift side.

Over the past few years, many government entities charged with mitigating coastal land loss have been turning to nature-based solutions such as planting mangrove forests, diverting sediment-laden water into wetlands, nurturing coral reefs, and abandoning "grey infrastructure" (e.g. Temmerman et al., 2013; Livingston et al., 2019; USACE, 2017). Seaweed farms that operate in coastal waters in China, Indonesia, and the Philippines provide jobs and help to attenuate waves while capturing and storing atmospheric carbon. Organizations such as the World Wildlife Foundation have sponsored programs such as Green Gravel Portugal to restore kelp (*Laminaria ochroleuca*) forests along the Portuguese coast. Pereira et al. (2019) described the life history of this native kelp in Portugal and conditions for its success. Bridges (2018) discussed the importance of aligning natural and engineering processes to collaboratively develop sustainable infrastructure. Combinations of intermittently spaced minimalist hard structures and natural wetlands, or mangroves offer intermediate approaches termed "living shorelines". In Australia today, the value of "millennia of experience" and understanding of Aboriginal people in regarding adaptations to changing environmental conditions is being increasingly acknowledged in adaptive planning (e.g. Matthews et al., 2023). The resilience-enhancing wisdom of Native Americans has also been recognized in the United States (Hutton & Allen, 2020). Nature-based solutions address coastal resilience through actions to protect, sustainably manage, and restore natural ecosystems. These approaches benefit people and nature at the same time.

Recovery in the wake of a natural or human-caused disaster depends on the combined strengths and capacities of engineers, scientists, individuals, and the organizations that comprise a community. Some additional resources are provided in Table 6.1. The starting point to recovery is knowing the risks is preparation (e.g., developing a plan to mitigate those risks). The risks are increasing owing to climate change. Communities might adapt to the adverse effects of climate in various ways, ranging from installation of early warning systems or planting drought-resistant crops. In extreme cases, relocation may be the last option, but this will require extensive engagement of threatened communities and convincing communications from well-trusted scientists.

Resilience involves societal, economic, and ecological considerations. Because all communities face hazards, applying state-of-the-art solutions to improving resilience is an imperative. Many communities are neither prepared to respond to emergency situations nor prepared to recover from the aftermath (e.g., Ma et al., 2023; Sledge & Thomas, 2019). For this reason, governments need to help improve resilience through responsive policy-making structures that promote the 4 C's (collaboration, communication, cooperation, and coordination) while reducing institutional barriers. Collaborative programs that include engineers, scientists, and the community might be designed to prevent short-term hazard events such as nuisance flooding from turning into a long-term disaster. Scientists from academia might work closely with industry and community stakeholders to implement research results for a particular community. Engineers can help a community to resist and adapt to flooding through a range of projects from the installation of nature-based solutions such as oyster reefs and living shorelines to the relocation of infrastructure to higher elevations. The integration of the complementary capabilities among diverse stakeholders is the key to planning and completing effective projects that improve resilience to coastal hazards.

Table 6.1 Selected tools supporting coastal resilience projects. Many of these tools provide information, mapping capabilities, and networking suggestions to help communities prepare, adapt, recover, and resist natural hazards

Tool	Proponent	URL
Climate change adaptation and resources for Natural Resource Management	Commonwealth Scientific and Industrial Research Organization (CSIRO), which is Australia's National Science Agency	https://adaptnrm.csiro.au/
Coastal resilience tools	TNC	https://coastalresilience.org/tools/
Climate Risk and Resilience Portal (ClimRR)	Argonne National Laboratory	https://climrr.anl.gov/
Digital coast	NOAA	https://www.coast.noaa.gov/digitalcoast/
Drought resilience self-assessment tool	Australian Government	https://www.drsat.com.au/
En-ROADS	Climate Interactive and Massachusetts Institute of Technology	https://en-roads.climateinteractive.org/scenario.html?v=24.4.0
Global change explorer	US EPA	https://www.epa.gov/gcx
Quick risk estimation tool	UN Office for Disaster Reduction	https://mcr2030.undrr.org/quick-risk-estimation-tool
Resilience booster tool	The World Bank	https://resiliencetool.worldbank.org
Resilient cities network	Partnership sponsored by The Rockefeller Foundation	https://resilientcitiesnetwork.org/climate-resilience/
Resilience rising	Consortium focused on resilience worldwide	https://resiliencerisingglobal.org/
Resilience tool	World Governments Summit	https://www.worldgovernmentsummit.org/observer/resilience-tool
Resilience Tools Wizard	US EPA	https://www.epa.gov/emergency-response-research/environmental-resilience-tools-wizard
U.S. Climate resilience toolkit	US Government	https://toolkit.climate.gov/

Numerous projects that include TNC and its partners are described online at https://coastalresilience.org/our-work/

References

Bridges, T. (2018). Engineering with nature [video]. U.S. Army Corps of Engineers. https://vimeo.com/589625794/5891f80bbd?share=copy

Hutton, N. S., & Allen, T. R. (2020). The role of traditional knowledge in coastal adaptation priorities: The Pamunkey Indian reservation. *Water, 12*(12), 3548. https://doi.org/10.3390/w12123548

Livingston, J., Woiwode, N., Bortman, M., McAfee, S., McLeod, K., Newkirk, S., & Murdock, S. (2019). Natural Infrastructure to mitigate inundation and coastal degradation. In L. D. Wright & C. R. Nichols (Eds.), *Tomorrow's coasts: Complex and impermanent. Coastal Research Library, 27.* Springer, Cham, Switzerland, pp. 167–189

Ma, C., Qirui, C., & Lv, Y. (2023). "One community at a time": Promoting community resilience in the face of natural hazards and public health challenges. *BMC Public Health, 23*, 2510. https://doi.org/10.1186/s12889-023-17458-x

Matthews, V., Vine, K., Atkinson, A.-R., Longman, J., Lee, G., Vardoulakis, S., & Mohamed, J. (2023). Justice, culture, and relationships: Australian Indigenous prescription for planetary health. *Science, 381*, 636–641. https://doi.org/10.1126/science.adh9949

Pereira, T. R., Azevedo, I. C., Oliveira, P., Silva, D. M., & Sousa-Pinto, I. (2019). Life history traits of Laminaria ochroleuca in Portugal: The range-center of its geographical distribution. *Aquatic Botany, 52*, 1–9. https://doi.org/10.1016/j.aquabot.2018.09.002

Sledge, D., & Thomas, H. F. (2019). From disaster response to community recovery: Nongovernmental entities, government, and public health. *American Journal of Public Health., 109*(3), 437–444. https://doi.org/10.2105/AJPH.2018.304895

Temmerman, S., Meire, P., Bouma, T., Herman, P., Ysebaert, T., & De Vriend, H. J. (2013). Ecosystem-based coastal defense in the face of global change. *Nature,504*, 79–83. https://doi.org/10.1038/nature12859

USACE. (2017). U.S. Army Corps of Engineers Resilience Initiative Roadmap, EP 1100-1-2. U. S. Department of the Army, Washington, D.C. https://www.publications.usace.army.mil/Portals/76/Publications/EngineerPamphlets/EP_1100-1-2.pdf?ver=2017-11-02-082317-943

Glossary[1]

Accretion Deposition of sediments along the beach or foreshore.

Adaptation A process where actions are adjusted to adapt to the impacts of the changing climate. These actions are solutions to the impacts of a changing climate.

Citizen scientist A member from the community who collects and analyzes data, e.g., environmental information, while participating in collaborative research with university and industry scientists.

Climate the long-term pattern of atmospheric and oceanic conditions at a location.

Coastal resilience Developing the ability to "bounce back" after hazardous events such as hurricanes, coastal storms, and flooding (NOS, 2024).

Consequence The impact of climate change can be identified as either negligible, marginal, critical, or catastrophic.

Glacial isostatic adjustment The rebound of land once burdened by ice-age glaciers. Earth's movements are monitored by networked Global Positioning System stations such as the Continuously Operating Reference Stations, European Terrestrial Reference System, and the Global Geodetic Observing System (see https://ggos.org/).

Intermediary A person of organization who mediates between collaborating partners. Also known as a broker or broker organization.

Inundation The total water level occurring on normally dry ground as a result of storm surge. The USGS implemented the Surge, Wave, and Tide Hydrodynamics (SWaTH) network to monitor and document the height, extent, and timing of storm surge (Verdi et al., 2017). NOAA provides an online Inundation Analysis Tool (see https://tidesandc urrents.noaa.gov/inundation/).

Erosion The loss of land or the removal of beach or dune sediments by wave action, tidal currents, wave currents, or drainage.

Exposure The degree to which people could be affected by coastal hazards.

---

[1] This glossary provides definitions of significant, common terms relating to coastal resilience.

C. Reid Nichols et al., *Integrated Coastal Resilience*, Synthesis Lectures on Ocean Systems Engineering, https://doi.org/10.1007/978-3-031-68153-0

General circulation models Mathematical models capable of representing physical processes of the atmosphere and ocean to simulate response of global climate to the increasing greenhouse gas emission (IPCC, 2013).

Marsh migration Marsh migration describes the process where marsh plants replace non-wetland coastal plants that dye along marsh borders which creep landward owing to sea level rise.

Mesoscale Weather patterns such as squall lines and flash floods with horizontal dimensions ranging from 5 to several hundred kilometers.

Mitigation Actions that prevent or slow the rate of climate change and contribute to lowering the effect of climate change factor.

Relative Sea Level Rise Changes in the height of sea water relative to land with respect to the average conditions over a reference period.

Risk A hazard or the chance of a loss.

Salt water front A diffuse boundary in the aquifer where fresh water and saltwater mix.

Scour The removal of granular bed material in the vicinity of coastal structures such as pilings owing to waves, tides, and currents.

Stakeholder Individuals or groups who have current and past experiences in preparing for, resisting, recovering, and adapting to climate variability and extremes.

Storm surge The abnormal rise of water generated by a storm, over and above the predicted astronomical tides.

Subsidence The gradual settling or sudden sinking of the Earth's surface owing to subsurface movement of earth materials.

Susceptibility the tendency that an area will undergo the effects hazardous processes such as dead zones, droughts, floods, harmful algal blooms, sea level rise, subsidence, etc.

Uncertainty While there are various types of uncertainty, when one thinks about resilience, uncertainty relates to the lack of sureness about predictive models. The uncertainties are due to an incomplete understanding of Earth processes and their interactions.

Vulnerability A combination of exposure, sensitivity and capacity to adapt to the changing climate.

Weather The state of the atmosphere or ocean at a given point in time and geographic location.

References

NOS. (2024). What is resilience? [video]. National Ocean Service. https://oceanservice.noaa.gov/facts/resilience.mp4

Verdi, R. J., Lotspeich, R. R., Robbins, J. C., Busciolano, R. J., Mullaney, J. R., Massey, A. J., Banks, W. S., Roland, M. A., Jenter, H. L., Peppler, M. C., Suro, T. P., Schubert, C. E., & Nardi, M.

R. (2017). The surge, wave, and tide hydrodynamics (SWaTH) network of the U.S. Geological Survey—Past and future implementation of storm-response monitoring, data collection, and data delivery. U.S. Geological Survey Circular 1431, United States Geological Survey. https://doi.org/10.3133/cir1431

IPCC. (2013). Climate change 2013: The physical science basis. contribution of working group i to the fifth assessment report of the intergovernmental panel on climate change [Stocker, T.F., D. Qin, G.-K. Plattner, M. Tignor, S.K. Allen, J. Boschung, A. Nauels, Y. Xia, V. Bex and P.M. Midgley (eds.)]. Cambridge University Press. https://www.ipcc.ch/site/assets/uploads/2017/09/WG1AR5_Frontmatter_FINAL.pdf